Springer Series in Materials Science

Volume 194

For further volumes:
http://www.springer.com/series/856

The Springer Series in Materials Science covers the complete spectrum of materials physics, including fundamental principles, physical properties, materials theory and design. Recognizing the increasing importance of materials science in future device technologies, the book titles in this series reflect the state-of-the-art in understanding and controlling the structure and properties of all important classes of materials.

S. V. Gupta

Viscometry for Liquids

Calibration of Viscometers

Springer

S. V. Gupta
Delhi
India

ISSN 0933-033X ISSN 2196-2812 (electronic)
ISBN 978-3-319-04857-4 ISBN 978-3-319-04858-1 (eBook)
DOI 10.1007/978-3-319-04858-1
Springer Cham Heidelberg New York Dordrecht London

Library of Congress Control Number: 2014932968

Printed on acid-free paper

Springer is part of Springer Science+Business Media (www.springer.com)

Dedicated to my wife Mrs. Prem Gupta
To my children, grand children
and great grand children
as They all inspire me to live longer

Preface

This book has been initiated by my former colleague, Mr. Anil Kumar Senior Principle Scientific Officer at National Physical Laboratory, who put to me some questions regarding the calibration of viscometers. I also realised that understanding of viscosity and viscous forces have been a subject matter of interest to research workers in many fields like medicine, oil industry, to physicists, chemists, engineers, fuel technologists and to rheologists.

Absolute measurement of viscosity of liquids even with 0.5 % accuracy is rather a tough and time-consuming job. On the other hand, measurement of viscosity relative to viscosity of some standard has been much easier and uncertainty involved is much less than 0.1 %. Hence it becomes necessary to have liquid of known viscosity. Easy availability of water of known characteristics and density to very high degree of accuracy makes it a correct choice to be the primary standard of viscosity. Therefore, the measurement of viscosity of water has been emphasised. Some scientists have spent a major part of their working life in viscosity measurement of water. For example, Scientists at NIST USA have spent some 20 years to establish the value of viscosity of water. The measurement of viscosity of water by capillary and oscillations viscometers has been discussed in Chap. 8 of the book.

Once the value of viscosity of water is established, the next valid step is to build the viscosity scale. This has been discussed in Chap. 2 of the book. Uncertainty propagated to the nth step of viscosity scale has also been derived. Corrections necessary to apply to viscometer constant of the viscometer when used at different temperatures with a different liquid in establishing the viscosity scale have been discussed in detail.

Keeping the popularity of capillary viscometers in mind turbulent and streamline flow has been given. Expression for kinematic viscosity in terms of efflux time and dimensions of the capillary viscometer is established. It has been observed that the kinetic energy correction given in different documents vary in the numerical factor which causes a lot of confusion. Different numerical factors used in the kinetic energy correction are due to use of incoherent units. This point has been amply discussed in Chap. 1.

The rotating and oscillating viscometers along with the viscometers used in specific fields have been discussed in detail. New trends based on modern physical principles like use of PZT crystal, Optical fibre shear waves and Love waves have been discussed to meet the inner hunger of the researchers in the field. Availability of commercial viscometers, advantage of specific type viscometers have been discussed for the convenience of the users.

It is my pleasant duty to thank my former colleagues at NPL such as Mr. Anil Kumar who along with Mrs. Reeta Gupta has helped me provide all the literature and scientific papers. I need to gratefully acknowledge the help of Dr. Ashok Kumar who has gone through my write-up and advised me on viscometers based on ultrasonic and shear waves PZT crystals. Nothing will be enough to thank Dr. Habil Claus E. Ascheron my Editor at Springer, without whose help and guidance it would have been impossible to get this book published.

Delhi, India, 14 October, 2013 S. V. Gupta

Contents

Chapter 1
Flow Through Capillary

Abstract With introductory remarks about laminar and turbulent flow, expression for rate of discharge of liquid for a laminar flow has been derived. The expression does contain kinetic energy term. Different confusing numerical factors appearing in various literatures in the kinetic energy term have been explained. SI units of dynamic and kinematic viscosities together with their corresponding CGS units have been given. Kinematic viscosity in terms of efflux time for given volume of liquid is given in terms of efflux time. Corrections applicable to viscometer constants to capillary viscometers due to various parameters have been derived. The parameters are gravity, buoyancy, thermal expansion of viscometer bulb, different temperatures of loading and use, surface tension, temperature coefficient of viscosity etc.

1.1 Types of Flow

1.1.1 Turbulent Flow

The type of fluid flow, in which local velocities and pressures that fluctuate randomly, is known as turbulent flow. In turbulent flow the speed of the fluid at a point is continuously changing both in magnitude and direction. The flow of wind and rivers is generally turbulent, even if the currents are gentle. The air or water swirls and eddies while its overall bulk moves along a specific direction.

Most kinds of fluid flow are turbulent, except for laminar flow at the leading edge of solids moving relative to fluids or extremely close to solid surfaces, such as the inside wall of a pipe, or in cases of fluids of high viscosity (relatively great sluggishness) flowing slowly through small channels. Common examples of turbulent flow are blood flow in arteries, oil transport in pipelines, lava flow, atmosphere and ocean currents, the flow through pumps and turbines, and the flow in boat wakes and around aircraft-wing tips. The velocity profile of different layers in turbulent flow is shown in Fig. 1.1.

S. V. Gupta, *Viscometry for Liquids*, Springer Series in Materials Science 194,
DOI: 10.1007/978-3-319-04858-1_1, © Springer International Publishing Switzerland 2014

Fig. 1.1 Turbulent flow

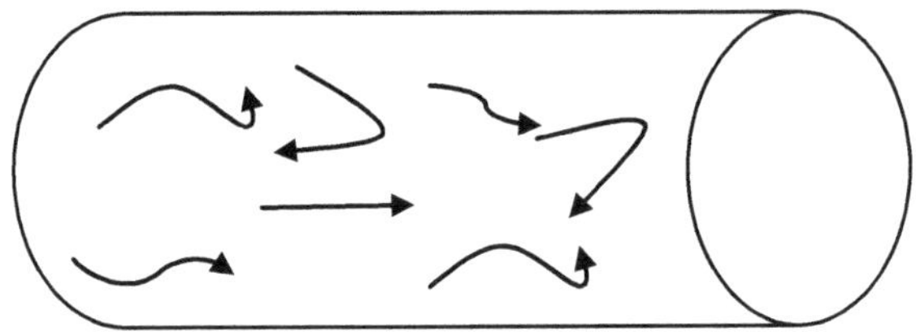

Fig. 1.2 Laminar flow

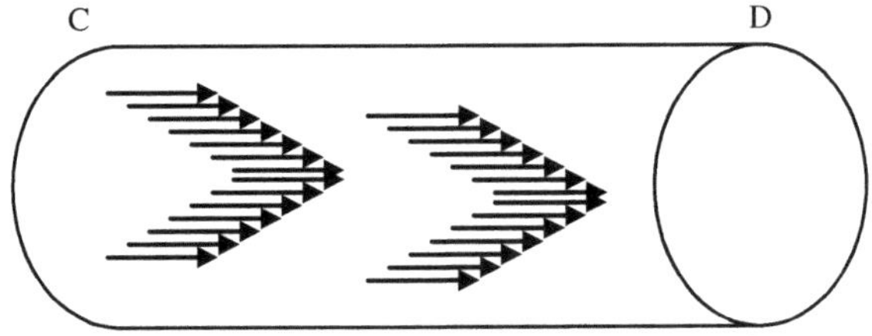

1.1.2 Laminar Flow

The laminar flow is the type of fluid flow in which the fluid moves in smooth paths or layers. All the fluid is flowing in the same direction. Fluid in contact of the solid stationary objects like the walls of a channel has zero velocity and velocity of consecutive layers increases away from the walls. In a closed channel say a cylindrical pipe the velocity of the layer of fluid along the axis of the pipe is maximum. The liquids of high viscosity moving in relatively small bore tube have the laminar flow. The velocity profile of different layers of liquids with laminar flow in a circular tube is shown in Fig. 1.2.

1.2 Motion in Laminar Flow

In a laminar flow, due to relative motion of layers with respect of each other, a resistive force acts between any two consecutive layers. This resistive force per unit area (shearing stress) divided by rate of change of velocity (shearing strain) is known as coefficient of dynamic viscosity or simply dynamic viscosity. This is normally represented by η.

Let us consider the laminar flow of a liquid along a fixed surface LM, Fig. 1.3. It is experimentally found that a layer CD at a distance $y + dy$ from LM flows faster than the layer AB distance y from the fixed surface LM. If the difference between the velocities of two layers is dv, the velocity gradient along y- the direction perpendicular to the direction of flow is $\frac{dv}{dy}$. Due to this relative motion of

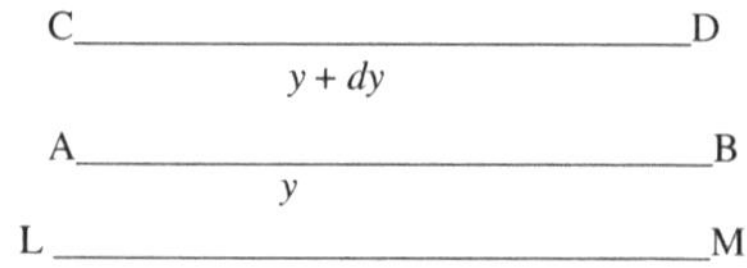

Fig. 1.3 Motion in laminar flow

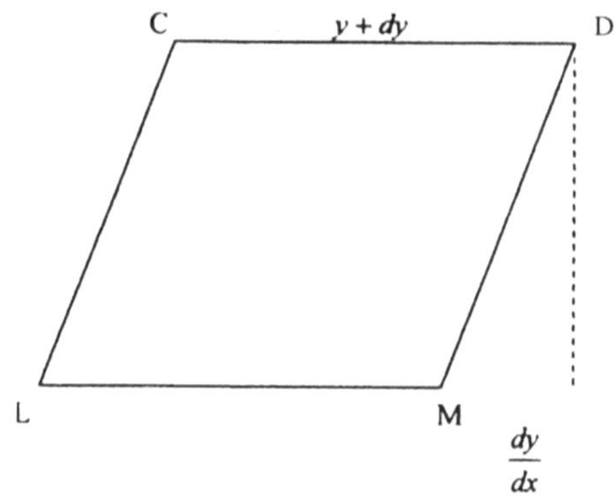

Fig. 1.4 Shearing of the flowing liquid

the layers, resistive force due to internal friction or viscosity acts and whole liquid will shear in the direction of laminar flow and it vertical section will look like as shown in the Fig. 1.4. If F is the force between any two consecutive layers having an area A, then shearing stress is $\frac{F}{A}$ and shearing strain is $\frac{dv}{dy}$. Hence for liquids flowing with the laminar flow, the coefficient of dynamic viscosity η is given as

$$\eta = \frac{F/A}{\frac{dv}{dy}} \tag{1.1}$$

Giving F the frictional force as

$$F = \eta A \frac{dv}{dy} \tag{1.2}$$

This is the Newton's Law of viscous flow for stream line (laminar) flow.

1.3 Unit of Dynamic Viscosity

To obtain the unit of dynamic viscosity (symbol η), write the units of quantities in (1.1), in which it has been defined.

$$\text{SI Unit of dynamic viscosity} = \frac{\text{N/m}^2}{(\text{m/s})/\text{m}} = \frac{\text{pascal}}{1/\text{s}} = \text{Pa}\cdot\text{s} \tag{1.3}$$

Similarly the unit of dynamic viscosity η expressed in terms of stress (F/A) and strain

$$\text{(velocity gradient) in CGS system} = \frac{\text{dyn/cm}^2}{(\text{cm/s})/\text{cm}} = \frac{\text{dyn·s}}{\text{cm}^2} = \text{poise} \quad (1.4)$$

The poise is the unit of dynamic velocity in CGS system.

Expressing the CGS units in terms of SI units in (1.4), we get the relation between CGS and SI units.

$$\text{poise} = \frac{10^{-5}\text{N·s}}{10^{-4}\text{m}^2} = 10^{-1}\text{Pa·s} = 0.1\ \text{Pa·s} \quad (1.5)$$

The symbol of CGS unit of dynamic viscosity poise is P, giving

$$1\ \text{P} = 0.1\ \text{Pa·s} \quad (1.6)$$

or

$$10^{-2}\,\text{P} = 1\ \text{cP} = 10^{-3}\text{Pa·s} = \text{mPa·s}$$

$$1\ \text{cP} = 1\ \text{mPa·s} \quad (1.7)$$

In the older literature, the values of dynamic viscosity of different fluids are given in terms of centi-poise symbol (cP). Hence from (1.7) the values of dynamic viscosities in cP would remain unchanged when expressed in millipascal second (mPa·s) a sub-multiple of SI unit.

1.4 Rate of Flow in a Capillary

Assuming that there is a laminar flow that means (1) pressure over any cross-section is constant, (2) there is no radial flow and (3) the liquid in contact of the wall of the circular tube is stationary, we can arrive at a theoretical formula for rate of flow of the liquid.

Referring to Fig. 1.5, let the velocity of the liquid layer at a distance r from the axis of the tube be v, and that of the layer $r + dr$ be $v - dv$, then the velocity gradient will be $-\frac{dv}{dr}$ giving tangential stress from (1.1) as

$$-\eta\frac{\text{dv}}{\text{dr}} \quad (1.8)$$

If the pressure difference between two points, distance L apart, is p, then the driving force on the liquid will be given as

$$\pi r^2 p \quad (1.9)$$

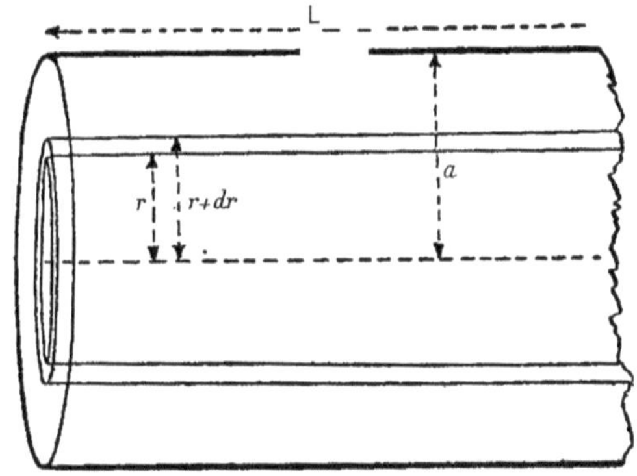

Fig. 1.5 Flow through a capillary

While the viscous force will be

$$-\eta \frac{dv}{dr} 2\pi rL \tag{1.10}$$

Here $2\pi rL$ is surface area of the cylindrical layer at a distance r from the axis. If this entire driving force is utilised in overcoming the viscous force, then

$$p\pi r^2 = -\eta \frac{dv}{dr} 2\pi rL \tag{1.11}$$

Giving

$$\frac{dv}{dr} = -\frac{rp}{2\eta L} \tag{1.12}$$

We know that v is zero at the wall i.e. at $r = a$. Integrating the (1.12) for r between 0 to a, we get

$$\mathrm{v} = \frac{p}{4\eta L}(a^2 - r^2) \tag{1.13}$$

The velocity profile across the cross-section is, therefore, a parabola.

Here $2\pi rdr$ is the cross-sectional area of the ring bounded by the two layers dr apart and v is the velocity so dQ the rate of flow of the liquid is the product of area of cross-section and velocity v of the liquid. Hence dQ is given by

$$dQ = 2\pi rvdr$$

Substituting the value of v from (1.13) and integrating it with respect of r for limits 0 to a, and, we get

$$Q = 2\pi \frac{p}{4\eta L} \int_0^a r(a^2 - r^2)dr$$

$$Q = \frac{p\pi a^4}{8\eta L} \tag{1.14}$$

If V is the volume of liquid flown in the time T then

$$\frac{V}{T} = Q = \frac{p\pi a^4}{8\eta L} \tag{1.15}$$

giving η as

$$\eta = \frac{p\pi a^4}{8VL} T. \tag{1.16}$$

1.4.1 Kinetic Energy Correction

In driving (1.11) it is assumed that entire pressure difference p is used for overcoming the viscous forces, however it is a fact that the liquid has acquired a certain velocity hence some pressure difference is required for imparting this velocity. The rate of kinetic energy acquired by the liquid is derived as follows:

Volume flow rate of the liquid between the two layers dr apart and at a distance r from the axis is

$$2\pi rvdr$$

If ρ is the density of the liquid then mass flow rate is given by

$$2\pi\rho rvdr$$

Hence rate of kinetic energy acquired by the liquid $\frac{1}{2}2\pi\rho rvdrv^2$

Using (1.13), total rate of kinetic energy K is given as

$$K = \int_0^a \pi\rho r\left(\frac{p}{4\eta L}\left(a^2 - r^2\right)\right)^3 dr$$

$$\text{Put} \quad a^2 - r^2 = z$$
$$-2r\,dr = dz$$

The limits of the variable z are a^2 to 0, giving us

$$\begin{aligned} K &= -\int_{a^2}^{0} \frac{\pi\rho}{2}\left(\frac{p}{4\eta L}\right)^3 z^3 dz = \frac{\pi\rho}{2}\left(\frac{p}{4\eta L}\right)^3 \frac{a^8}{4} \\ &= \left(\frac{\pi p a^4}{8\eta L}\right)^3 \frac{\rho}{\pi^2 a^4} = \frac{Q^3\rho}{\pi^2 a^4} \end{aligned} \tag{1.17}$$

The energy spent per second in moving the liquid through capillary is pQ. If p_1 is total external pressure then $p_1 Q$ is total rate of loss of energy giving us

$$p_1 Q = pQ + \frac{Q^3 \rho}{\pi^2 a^4} \tag{1.18}$$

Giving

$$p = p_1 - \frac{Q^2 \rho}{\pi^2 a^4} \tag{1.19}$$

Substituting this value of p in (1.16) and replacing Q by $\frac{V}{T}$, we get

$$\begin{aligned} \eta &= \frac{\pi a^4}{8VL} T \left[p_1 - \frac{V^2 \rho}{T^2 \pi^2 a^4} \right] \\ \eta &= \frac{\pi p_1 a^4}{8LV} T - \frac{V\rho}{8\pi L} \frac{1}{T} \end{aligned} \tag{1.20}$$

In driving (1.20), the energy required in only for the axial flow, however there is some radial flow also, so we use a coefficient say m to compensate for radial kinetic energy. Hence more complete equation is

$$\eta = \frac{\pi p_1 a^4}{8LV} T - \frac{V\rho}{8\pi L} \frac{m}{T} \tag{1.21}$$

In case of a capillary viscometer

$$p_1 = h\rho(1 - \sigma/\rho)g \tag{1.22}$$

h being the fall of the menisci when fixed volume of the liquid has flown through the capillary viscometer.

Substituting the value of p_l from (1.22) and dividing both sides of (1.21) by ρ, we get

$$\frac{\eta}{\rho} = \frac{\pi h(1 - \sigma/\rho)ga^4}{8VL} T - \frac{mV}{8\pi L} \frac{1}{T} \tag{1.23}$$

For a given viscometer $\frac{V}{8\pi L}$ is constant, so writing B for it, and C for $\frac{\pi h a^4 (1-\sigma/\rho)g}{8LV}$, (1.23) may be written as

$$\frac{\eta}{\rho} = CT - \frac{mB}{T} \tag{1.24}$$

In earlier days m was taken as constant however Cannon and Manning [1] proved that m is not constant. For a trumpet shaped capillary ends m can be represented best as

$$m = 0.037(\text{Re})^{1/2} \tag{1.25}$$

Multiplying factor is 0.037 for trumpet shape ends of the capillary tube, for square cut ends this factor is more than 0.037. However capillary tubes with square cut ends are not recommended as there is no specific advantage of the use of such tubes but with a distinct disadvantage of having more kinetic energy correction.

Here Re is the Reynolds number given by

$$Re = \frac{\rho v L_1}{\eta} \tag{1.26}$$

Here

ρ is the density of the liquid
v is the average velocity
L_1 is distance over which velocity is averaged out and
η is the dynamic viscosity.

In the case of flow through a capillary, average velocity $v = \frac{Q}{\pi D^2/4} = \frac{4V}{\pi D^2 T}$ and the L_1 in (1.26) is the diameter D as the average is taken over the diameter.

Substituting the values of v and L_1 in the definition of Reynolds number, (1.25) becomes

$$m = 0.037\left[\frac{4VD\rho}{\pi D^2 T\eta}\right]^{1/2} \tag{1.27}$$

For a properly made viscometer the second term in (1.24) is only 3 % of the first term, hence replacing η/ρ by CT in (1.27) will cause an error of not more than 0.09 %.

Hence (1.27) becomes

$$m = 0.074\left[\frac{V}{\pi CDT^2}\right]^{1/2} \tag{1.28}$$

The second term in (1.24) becomes

$$\begin{aligned}\frac{mB}{T} &= \frac{1}{T}0.074\left[\frac{V}{\pi CD}\right]^{1/2}\frac{1}{T}\frac{V}{8\pi L}\\ &= \frac{0.074V^{3/2}}{8\pi^{3/2}L(CD)^{1/2}}\frac{1}{T^2}\end{aligned} \tag{1.29}$$

$$\frac{mB}{T} = \frac{0.00166V^{3/2}}{L(CD)^2}\cdot\frac{1}{T^2} \tag{1.30}$$

Substituting the value of $\frac{mB}{T}$ from (1.29) in (1.24), we get

$$\frac{\eta}{\rho} = CT - \frac{E}{T^2} \tag{1.31}$$

Here

$$E = \frac{0.00166V^{3/2}}{L(CD)^{1/2}} \tag{1.32}$$

But $\frac{\eta}{\rho}$ is the kinematic viscosity of the liquid with symbol v. So the v can be expressed as

$$v = CT - \frac{E}{T^2} \tag{1.33}$$

Here

$$C = \frac{\pi h a^4 (1 - \sigma/\rho) g}{8LV} \text{ and E} = \frac{0.00166V^{3/2}}{L(CD)^{1/2}} \tag{1.34}$$

It can be easily seen that the dimension of each term on right hand side of (1.33) is (Length)^2/time. Hence SI unit of v is m^2 s^{-1}, of C is m^2 s^{-2} and that of E is m^2 s.

It has been noticed that the numerical factor of E has been given different in various national and International documents. The reason is the use of mixed system of units. The value of the multiplying factor for E in (1.32) is 0.00166 when all the quantities are expressed in coherent units of measurement. Say if V is in m^3, L and D should be in m and C in m^2/s^2 then unit of E is m^{2}s and that of v is m^2/s. Similarly if we take V in cm^3, L and D in cm and C in cm^2/s^2 then unit of E shall be cm^2 s and that v of cm^2/s.

However if V is taken in cm^3, L and D are in cm but C in mm^2/s^2 (mixed system of units) then the factor in E becomes

$$\begin{aligned} \frac{0.00166(10^{-6}\text{m}^3)^{3/2}}{10^{-2}\text{m}(10^{-6}\text{m}^2\,\text{s}^{-2}10^{-2}\text{m})^{1/2}} &= 0.00166\frac{10^{-9}}{10^{-6}}\text{m}^2\,\text{s} \\ &= 0.00166 \cdot 10^{-3} \cdot 10^{6}\text{mm}^2\,\text{s} \\ &= 1.66\ \text{mm}^2\,\text{s} \end{aligned} \tag{1.35}$$

This is the value of the factor given by Fujita et al. in Metrologia paper 2009 [2]. If V is taken in cm^3, L and D are in mm and C in mm^2/s^2, then the factor in E is

$$\begin{aligned} \frac{0.00166(10^{-6}\text{m}^3)^{3/2}}{10^{-3}\text{m}(10^{-6}\text{m}^2\,\text{s}^{-2}\cdot 10^{-3}\text{m})^{1/2}} &= \frac{0.00166 \cdot 10^{-9}\text{m}^{9/2}}{10^{-3}\text{m} \cdot 10^{-9/2}\text{m}^{3/2}\,\text{s}^{-1}} \\ &= 0.00166 \cdot 10^{-3/2}\text{m}^2\,\text{s} \\ &= 0.00166 \cdot 10^{-3/2} \cdot 10^{6}\text{mm}^2\,\text{s} \\ &= 52.5\ \text{mm}^2\,\text{s} \end{aligned} \tag{1.36}$$

ASTM D446-07 [3] has given this value of the factor in E.

1.4.2 End Correction

An additional correction is required to account for the peculiar phenomena at the ends of the tube. It is the work done against the viscous forces in the re-arrangement of velocity distribution. The effect of this is the resistance offered in converging at the entrance and diverging of steam at the exit ends of the tube. The effect is proportional to the radius of the tube and is expressed as a hypothetical addition to the length. The effective length L of the tube then becomes $L + na$. For further details of end corrections (change in effective length of the flow tube) one may consult [4]. So finally (1.21) becomes:

$$\eta = \frac{\pi a^4 p_1}{8V(L+na)}T - \frac{m\rho V}{8\pi(L+na)T}. \tag{1.37}$$

1.5 Units of Kinematic Viscosity

By definition the kinematic viscosity of a fluid is the ratio of dynamic viscosity to its density.

Units of dynamic viscosity is

$$\text{Pa}\cdot\text{s} = \text{kg}\cdot\text{m}^{-1}\cdot\text{s}^{-2}\cdot\text{s} = \text{kg}\cdot\text{m}^{-1}\cdot\text{s}^{-1}$$

$$\text{Unit of kinematic viscosity} = \frac{\text{kg}\cdot\text{m}^{-1}\cdot\text{s}^{-1}}{\text{kg}\cdot\text{m}^{-3}} = \text{m}^2\cdot\text{s}^{-1} \tag{1.38A}$$

But metre square per second is quite big unit so it's sub-multiple namely mm^2/s is used. For example 1.0034 mm^2/s is the value of kinematic viscosity of double distilled water at 20 °C.

$$\text{m}^2/\text{s} = 10^6\text{mm}^2/\text{s} \tag{1.38B}$$

CGS unit of kinematic viscosity is stoke (St) = $[\{\text{dyne}\cdot\text{cm}^{-2}\}\cdot\text{s}]/\text{g}\cdot\text{cm}^{-3}$

Expressing the right hand side in SI units, we get

$$\begin{aligned} &= \left[10^{-5}\text{N}/10^{-4}\text{m}^2\cdot\text{s}\right]/10^3\text{kg}/\text{m}^3 \\ &= 10^{-4}\text{m}^2/\text{s} = 10^2\text{mm}^2/\text{s} \end{aligned}$$

Giving us

$$1 \text{ centi} - \text{stoke (cSt)} = 1 \text{ mm}^2/\text{s} \tag{1.39}$$

The values of kinematic viscosities expressed in centi-stokes remain unchanged when quoted in mm^2s^{-1}.

1.6 Corrections to C Due to Various Parameters

Here we see that C the so called viscometer constant depends upon

1. the local acceleration due to gravity
2. the buoyancy:- densities of liquids and air used
3. the effective value of h which depends on several factors such as the difference in rise of liquid due to different surface tension of the liquids used and the difference in temperatures at which the viscometer was calibrated and used.
4. the volume of liquid V which in turn depends upon the temperature and expansion coefficients of liquid and glass used and essentially affect the effective pressure head h.

1.6.1 Correction Due to Gravity

If a viscometer is calibrated at a place with g_c as the acceleration due to gravity and is used at a place having acceleration due to gravity as g_u then the values of C at the places of calibration and use respectively are given as

$$C_c = \frac{\pi a^4(1 - \sigma/\rho)g_c}{8LV} \tag{1.40}$$

$$C_u = \frac{\pi a^4(1 - \sigma/\rho)g_u}{8LV} \tag{1.41}$$

Dividing (1.41) by (1.40), we get

$$C_u = C_c \frac{g_u}{g_c} \tag{1.42}$$

1.6.2 Buoyancy Correction

The c_b is air buoyancy correction on liquids of different density at the time of calibration and use. It is relative to calibration constant of viscometer C_c.

The viscometer constant C determined by calibrating the viscometer, using the standard oil of density ρ_c and with air density σ_c, is expressed as

$$C_c = \frac{\pi a^4(1 - \sigma_c/\rho_c)g_c}{8LV} \tag{1.43}$$

Similarly when the same viscometer is used to determine the viscosity of an oil of density ρ_u when air density is σ_u, C_u will be expressed as

$$C_{\mathrm{u}} = \frac{\pi a^4 (1 - \sigma_u/\rho_u) g_c}{8LV} \tag{1.44}$$

Dividing the two, we get

$$\frac{C_u}{C_c} = \frac{(1 - \sigma_u/\rho_u)}{(1 - \sigma_c/\rho_c)} = (1 - \sigma_u/\rho_u + \sigma_c/\rho_c)$$

$$C_{\mathrm{u}} = C_{\mathrm{c}}\{1 + (\sigma_c/\rho_c - \sigma_u/\rho_u)\} = C_{\mathrm{c}}(1 + c_{\mathrm{b}}) \tag{1.45}$$

Giving c_{b} the buoyancy correction relative to C_{c} as

$$c_{\mathrm{b}} = (\sigma_c/\rho_c - \sigma_u/\rho_u). \tag{1.46}$$

1.6.3 Correction Due to Thermal Expansion of Viscometer Bulb

The c_{e} is the correction for the effect of thermal expansion of the viscometer due to different temperatures at which the viscometer was calibrated and used for measurement of viscosity of a liquid.

Let volume of the timing bulb be V_{o} at °C then volume V_{c} at the temperature of calibration t_{c} will be

$$V_{\mathrm{c}} = V_{\mathrm{o}}(1 + 3\alpha t_{\mathrm{c}}) \tag{1.47}$$

Here α is the coefficient of linear expansion of the material of the viscometer. Similarly V_{u} at the temperature of use (measurement) t_{m} is given by

$$V_{\mathrm{u}} = V_{\mathrm{o}}(1 + 3\alpha t_{\mathrm{m}}) \tag{1.48}$$

Now viscometer constants at the temperatures of calibration and use respectively are:

$$C_c = \frac{\pi a^4 (1 - \sigma/\rho) g_c}{8LV_c} \tag{1.49}$$

$$C_{\mathrm{u}} = \frac{\pi a^4 (1 - \sigma/\rho) h g_c}{8LV_u} \tag{1.50}$$

Substituting the value of V_{u} from (1.48) in (1.50) and dividing (1.50) by (1.49) we get

$$C_u = C_c \frac{(1 + 3\alpha t_c)}{(1 + 3\alpha t_m)} \tag{1.51}$$

or

$$C_u = C_c\{1 - 3\alpha(t_m - t_c)\} \tag{1.52}$$

The correction c_e due to thermal expansion of bulb relative to C_c is given by

$$c_e = -3\alpha(t_m - t_c). \tag{1.53}$$

1.6.4 Correction to C Due to Different Temperatures of Loading and Use

Some viscometers like Cannon–Fenske routine, Pinkevitch, Cannon-Manning semi-micro and Cannon–Fenske opaque are charged (loaded) at room temperature but are calibrated or used at another fixed temperature of bath. The temperature of bath, depending upon the kinematic viscosity of the liquid under-measurement, may be any thing between −30 and 100 °C. As the capillary extends on both sides of the two fixed marks, a small change in volume will have larger change in the pressure head h. The change in volume of the under-measurement liquid will cause a change in the effective pressure head h.

If the viscometer is loaded with volume V_f at temperature t_f and measurement is carried out at a different temperature t_m then a correction c_f for thermal expansion is to be applied. If t_f is higher than t_m, then V_f the volume of liquid will be reduced on attaining the bath temperature and also the capacity of the timing bulb will reduce. The difference in two reductions will be the net reduction in volume. The net reduction in volume will affect the effective pressure head h. The net reduction divided by the effective area of the capillary at the timing mark will be the reduction in the effective pressure head h.

We know the mass of the loaded sample remains unchanged, hence

$$V_f\rho_f = V_m\rho_m$$

Giving

$$V_m = V_f\frac{\rho_f}{\rho_m} \tag{1.54}$$

The change in volume of the liquid, therefore, is

$$V_m - V_f = V_f\left[\frac{\rho_f}{\rho_m} - 1\right] \tag{1.55}$$

The reduction in capacity v_c of the viscometer is given by

$$v_c = 3V_f\alpha(t_m - t_f) \tag{1.56}$$

$$\text{Net change in volume } = V_\mathrm{f}\left[\frac{\rho_f}{\rho_m} - 1\right] - 3\alpha V_\mathrm{f}(t_\mathrm{m} - t_\mathrm{f}) \tag{1.57}$$

But reduction in h is net reduction in volume divided by the area of cross-section at the lower timing mark, giving us

$$\begin{aligned}\text{Reduction in } h &= \frac{V_f\{(\rho_f/\rho_m - 1) - 3\alpha(t_m - t_f)\}}{\pi a^2} \\ &= h_\mathrm{c} - h_\mathrm{u}\end{aligned} \tag{1.58}$$

Now

$$C_\mathrm{c} = \frac{\pi a^4(1 - \sigma/\rho)h_c g_c}{8LV_u} \tag{1.59}$$

And

$$C_\mathrm{u} = \frac{\pi a^4(1 - \sigma/\rho)h_u g_c}{8LV_u} \tag{1.60}$$

Subtracting

$$\mathrm{C_c} - \mathrm{C_u} = \frac{\pi a^4(1 - \sigma/\rho)g_c(h_c - h_u)}{8LV_u} \tag{1.61}$$

Giving

$$C_u = C_c - C_c\frac{(h_c - h_u)}{h_c} \tag{1.62}$$

$$C_u = C_c(1 + c_f) \tag{1.63}$$

From (1.58) and (1.63) we get

$$\mathrm{c_f} = \frac{V_f\{(\rho_f/\rho_m - 1) - 3\alpha(t_m - t_f)\}}{\pi a^2 h_c} \tag{1.64}$$

Here V_f, a and h should be measured in coherent units, otherwise an appropriate factor will appear. If h and a are expressed in millimetres the V_f should also be expressed in cubic millimetres. Normally, however, a and h are taken in millimetres and volume V_f is taken in centimetre cube then new value of $\mathrm{c_f}$ is multiplied by 1000 giving us

$$c_\mathrm{f} = 1000\frac{V_f\{(\rho_f/\rho_m - 1) - 3\alpha(t_m - t_f)\}}{\pi a^2 h} \tag{1.65}$$

Normally instead of radius a, the diameter d is measured hence a may be replaced by $d/2$ giving us

$$c_f = 4000\frac{V_f\{(\rho_f/\rho_m - 1) - 3\alpha(t_m - t_f)\}}{\pi d^2 h} \tag{1.66}$$

This is the expression used in ASTM D446-07. This correction is applied to the viscometer constant at temperatures other than the calibration temperature for Cannon–Fenske routine, Pinkevitch and Cannon-Manning semi-micro viscometers. The correction for Cannon–Fenske opaque (Reverse flow) viscometer given in (1.66) will be negative.

1.6.5 Correction to C Due to Change in Surface Tension

If the two menisci have different average radii and the surface tension of liquid under measurement differs substantially from the liquid by which viscometer was calibrated then a surface tension correction c_s relative to C_c is necessary.

We know that

$$\delta h \text{ the rise of liquid due to surface tension } = \frac{2\Gamma}{\rho g r} \tag{1.67}$$

Here Γ is surface tension, ρ is density of the liquid, g is acceleration due to gravity and r is radius of the capillary at the point of rise in liquid.

Let r_u and r_L be the radii of the timing bulb at the upper and lower graduation mark respectively; Γ_c be surface tension and ρ_c be the density of liquid used at the time of calibration.

Rise of liquid at the upper graduation mark δh_u will be

$$\delta h_u = \frac{2\Gamma_c}{\rho_c g r_u} \tag{1.68}$$

Similarly rise due to surface tension at the lower graduation mark δh_L is given by

$$\delta h_L = \frac{2\Gamma_c}{\rho_c g r_L} \tag{1.69}$$

The net change in effective height between the two graduation marks δh_c is given as

$$\delta h_c = \delta h_u - \delta h_L$$

or

$$\delta h_c = \frac{2\Gamma_c}{\rho_c g}\left(\frac{1}{r_u} - \frac{1}{r_L}\right) \tag{1.70}$$

Similarly if a different liquid of density ρ_t having surface tension Γ_t is used then, net change in effective height δh_{uu} between the two graduation marks is given as

$$\delta h_{uu} = (\delta h_u - \delta h_L)_{use}$$

or

$$\delta h_{uu} = \frac{2\Gamma_t}{\rho_t g}\left(\frac{1}{r_u} - \frac{1}{r_L}\right) \tag{1.71}$$

But

$$C_c = \frac{\pi a^4 (1 - \sigma/\rho) g_c}{8LV_u}(h_c + \delta h_c) \tag{1.72}$$

And

$$C_u = \frac{\pi a^4 (1 - \sigma/\rho) g_c}{8LV_u} \cdot (h_c + \delta h_{uu}) \tag{1.73}$$

Subtracting

$$C_u - C_c = \frac{\pi a^4 (1 - \sigma/\rho) g_c}{8LV_u}(\delta h_{uu} - \delta h_c) \tag{1.74}$$

$$C_u - C_c = C_c \frac{(\delta h_{uu} - \delta h_c)}{h_c} \tag{1.75}$$

Or

$$C_u = C_c\left[1 + \frac{(\delta h_{uu} - \delta h_c)}{h_c}\right] \tag{1.76}$$

$$C_u = C_c[1 + c_s)] \tag{1.77}$$

But $\frac{(\delta h_u - \delta h_c)}{h_c} = \delta h_{eff}$, which from (1.70) and (1.71) is given as

$$\delta h_{eff} = \frac{2}{gh_c}\left(\frac{1}{r_u} - \frac{1}{r_L}\right)\left(\frac{\Gamma_c}{\rho_c} - \frac{\Gamma_t}{\rho_t}\right) \tag{1.78}$$

Hence

$$c_s = \frac{2}{gh_c}\left(\frac{1}{r_u} - \frac{1}{r_L}\right)\left(\frac{\Gamma_c}{\rho_c} - \frac{\Gamma_t}{\rho_t}\right) \tag{1.79}$$

Here all parameters like g, h, r_u, r_L, Γ_c Γ_t, ρ_c and ρ_t must be expressed in SI units or in a set of coherent units.

1.6.6 Temperature Correction to Kinematic Viscosity

The kinematic viscosity of all liquids is temperature dependent. Hence there is a correction c_t due to the actual measurement temperature and the standard temperature at which viscosity value is stated. If α_v is coefficient of kinematic viscosity of the liquid under test and t_m and t_s are the measurement and standard temperatures then c_t is given as

$$c_t = \alpha_v(t_m - t_s). \tag{1.80}$$

References

1. Cannon MR, Manning RE (1960) The kinetic energy correction and a new viscometer. Anal Chem 32:355–360
2. Fujita Y, Kurano Y, Fujii K (2009) Evaluation of uncertainty in viscosity measurements by capillary master viscometers. Metrologia 46:237–248
3. ASTM D446-07 (2007) Standard Specifications and operating instructions for glass capillary kinematic viscometers
4. Barr G (1931) A monograph of viscometry, Chapter II. Humphrey Milford, New York

Chapter 2
Kinematic Viscosity Scale and Uncertainty

Abstract Step up procedure, starting with primary standard (water), for establishing kinematic viscosity scale in terms of standard viscometers and oils has been explained in detail. Measurement system of viscosity at three levels has been discussed. Details of master capillary viscometers of different viscometer constants are prescribed. Uncertainty of measurement and its propagation from one step to another has been discussed. Standard uncertainty of the viscosity of oil and viscometer constant used in the nth step of the step up procedure has been derived.

2.1 Primary Standard

For kinematic viscosity determinations, distilled water at 20 °C is taken as the primary standard. The value of kinematic viscosity of water at 20 °C has been adopted as 1.0034 mm^2/s [1, 2].

2.2 Establishing a Viscosity Scale

Establishing a viscosity scale means to obtain a set of, viscometers of known viscometer constants and standard oils of known viscosity. This chapter is essentially based on the paper by the author in MAPAN [3].

Step 1

Starting from the known viscosity of water, two master viscometers capable of measuring viscosity of liquids in the range 1–3 mm^2/s and having viscosity constants in the range of 0.001–0.003 mm^2/s^2 are calibrated at 20 °C. Aforesaid range of viscosity constants ensure that time of flow (efflux time) is more than 300 s but <1000 s. The efflux times for the viscometer should repeat within 0.1 %.

S. V. Gupta, *Viscometry for Liquids*, Springer Series in Materials Science 194,
DOI: 10.1007/978-3-319-04858-1_2,

The average of the efflux time T_1 is taken to calculate the viscometer constant C_1 from the following relation:

$$C_1 = 1.0034/T_1 \, \mathrm{mm}^2/\mathrm{s}^2 \tag{2.1}$$

The viscometer constant of the other master viscometer is similarly calculated.

These master viscometers are used to determine the kinematic viscosity of two standard oils of similar viscosities at 40 °C. Corrections, as described in Chap. 1, are applied for buoyancy and wherever necessary, for temperature and surface tension. Repeatability of efflux time T_2 should be ±0.1 %. The η_1 viscosity of oil is calculated from the equation

$$h_1 = C_1 T_2. \tag{2.2}$$

The average, of kinematic viscosities determined by each viscometer, is taken as the viscosity of the oil under-test, provided the viscosity values agree within 0.1 %.

The whole exercise is repeated for the other oil of similar viscosity.

Step 2

The oils whose viscosity have been measured in Step 1 are taken as standards of known viscosity for Step 2 and are used to calibrate another set of two master viscometers having the constants in the range of 0.003–0.009 mm^2/s^2 at 40 °C using the equation

$$C_2 = h_1/T_3 \tag{2.3}$$

These master viscometers are used to determine the kinematic viscosity of another set of standard oils whose viscosity is in the range of 9–12 mm^2/s.

From above one can see that starting point of viscosity scale is double distilled water, in each subsequent step the oil whose viscosity is measured in the previous step is used to calibrate the viscometer and the viscometer so calibrated is used to measure the viscosity of the standard oil of three times the viscosity of the starting oil.

Step 3

This step up process is repeated by using viscometers of gradually increasing viscometer constants and oils of higher viscosity, till the required range of viscosity is achieved. It may be noticed that at each stage there are two calibrated viscometers or two oils of known viscosity. This ensures lesser uncertainty. The whole process of building up the viscosity sale is shown in Fig. 2.1.

In Fig. 2.1, vertical lines indicate liquids and slant lines represent viscometers. First vertical line represents water. Here we see that each viscometer is used twice, first to be calibrated by the liquid of known viscosity and second time to measure the viscosity of a liquid of higher range. Similarly, each of the 2 oils is used twice, first its viscosity is measured by viscometer and then the oil is used for calibrating the viscometer of next higher range.

Fig. 2.1 Viscosity Scale

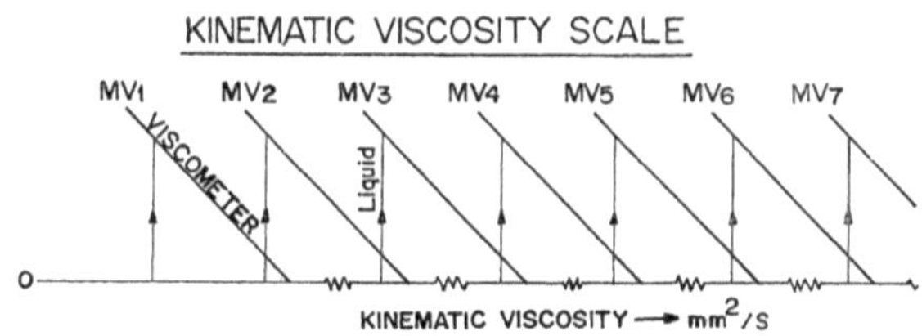

2.3 Viscosity Measurement System

2.3.1 At Level I

The viscosity measurement system is pictorially shown in Fig. 2.2. With the help of master viscometer MV_1, which has already been calibrated with water- the primary standard, kinematic viscosity of reference oil standard RL_1 is determined. RL_1 is used to calibrate master viscometer MV_2. MV_2 is used to determine the viscosity of RL_2. This process continues till the required viscosity range is reached.

2.3.2 At Level II

Reference liquids (oil standards) are used to calibrate level II or secondary viscometers (SV). Referring to Fig. 2.2, two references liquids RL_1 and RL_2 are used to calibrate a secondary viscometer. RL_1 has the viscosity close to the starting range of the viscometer SV and RL_2 has the viscosity towards the final range of the viscometer.

2.3.3 At Level III

Secondary viscometers are used to measure the kinematic viscosity of other oils and liquids.

2.4 Equipment Required

The Sects. 2.4 and 2.5 are essentially based on ASTM D2162-06 [2].

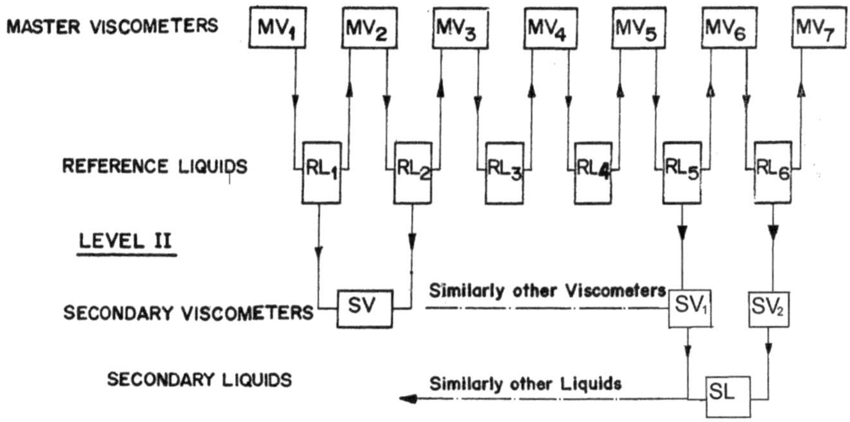

Fig. 2.2 Measurement system for kinematic viscosity

2.4.1 Master Viscometers

The viscometers used for the purpose are U tube Cannon or Ubbelohde type. The working length of the capillary of a master viscometer is 5–10 times that of the ordinary viscometer. This ensures stable stream line (laminar) flow and higher efflux time.

2.4.1.1 Cannon U Tube Master Viscometer

A typical Cannon U tube Master Viscometer is shown in Fig. 2.3.

The values of various labels may differ from one standard specification to another. For example the values of various symbols as per ASTM D 2162 are as follows:

OD1 = 12–13 mm; OD2 = 7–8 mm; OD3 = 8–9 mm; A1 = 100–125 mm, A3 = 35–40 mm; A4 = 25–35 mm; A5 = 30 ± 1 mm; A6 = 20–30 mm; A7 = 620–640 mm; Length of the capillary tube CL should be 445–450 mm, Bulb volume BV or BV1 may be 30–3.5 cm^3.

2.4.1.2 Dimensions of Master Viscometer

The essential dimensions of Cannon master viscometers used at the National Institute of Metrology Japan [3] are given in the Table 2.1.

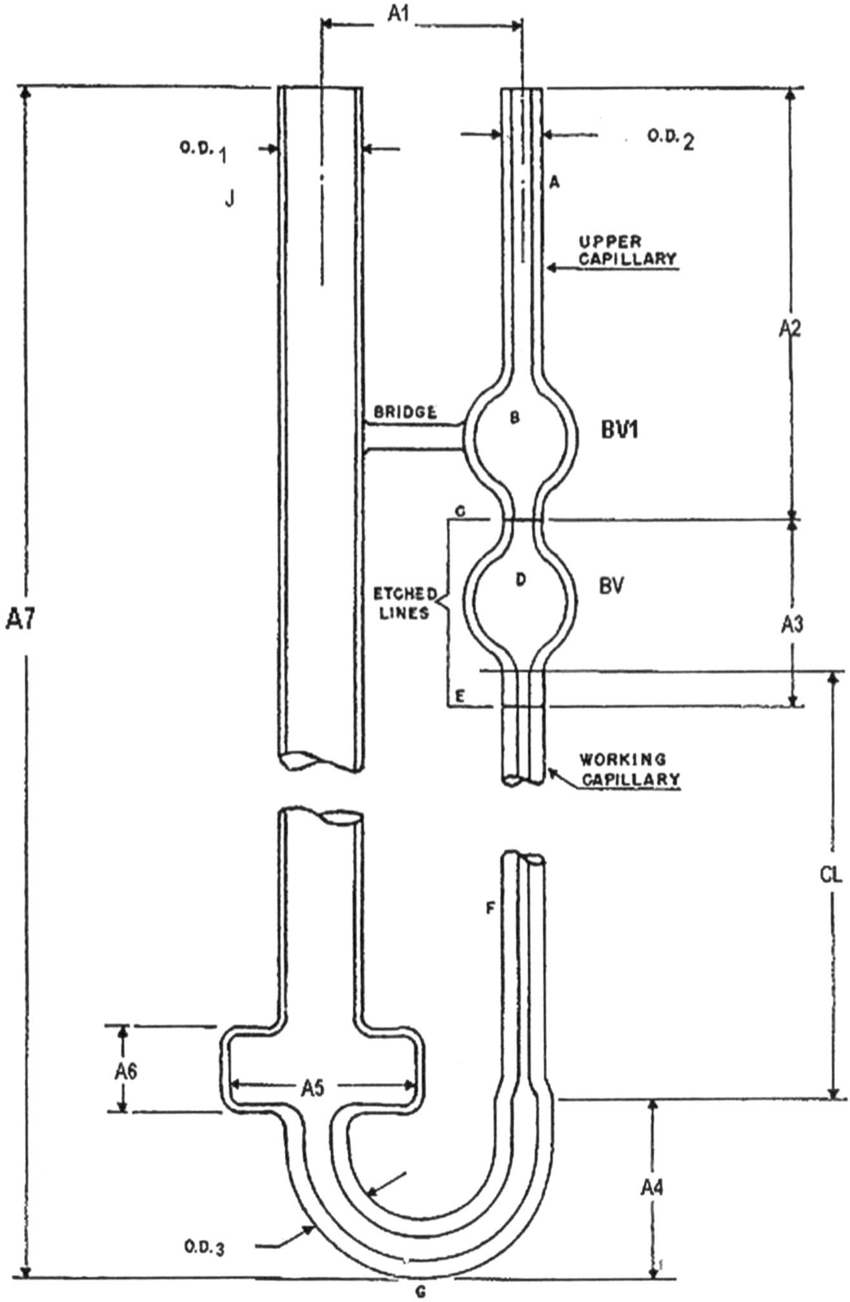

Fig. 2.3 Cannon master viscometer

Table 2.1 Dimensions of standard viscometers used in NMIJ

Viscometer no.	Viscometer constant (mm^2/s^2)	Capillary radius (mm)	Capillary length (mm)	Volume timing bulb (cm^3)	Mean head (mm)	Filling volume (cm^3)	Reservoir radius (cm^3)
1-11	0.00054	0.018	40.0	7.1	41.1	15.2	1.5
2-1	0.0035	0.024	40.0	3.1	41.5	11.1	1.5
3-1	0.0200	0.035	40.0	3.1	41.0	11.3	1.5
4-1	0.1000	0.053	40.0	3.1	41.2	11.3	1.5
5-1	0.42	0.079	40.0	3.1	41.6	11.0	1.4
6-1	2.7	0.14	40.0	5.0	41.3	16.6	1.5
7-1	12	0.23	40.0	9.9	41.1	35.3	1.6
8-1	33	0.32	40.0	9.8	41.3	42.7	2.2
9-1	145	0.42	40.0	9.3	41.3	54.3	2.3

2.4.1.3 Ubbelohde Master Viscometer

Another viscometer suitable for use as master viscometer is Ubbelohde type. A typical Ubbelohde master viscometer is shown in Fig. 2.4.

The values of various labels may differ from one standard specification to another. For example the values of various symbols as per ASTM D 2162 are as follows:

OD1 = 12–13; OD2 = 6–7 mm; OD3 = 17–19; OD4 = 9–10 mm; OD5 = 23–25 mm; OD6 = 6–12 mm; OD7 = 23.5; A1 = 22.5; A2 = 19, A3 = 70; A4 = 565 ± 10 mm; A5 = 400 ± 5 mm; A6 = 25 ± 2 mm; Bulb volume BV = 5.1 ± 1 cm^3.

2.4.2 Thermometers

Thermometers for kinematic viscosity measurements are of shorter range from 18.5 to 21.5 °C for 20 °C bath and 38.5–41.5 °C for 40 °C bath, each graduated at 0.01 °C intervals. Thermometers should conform to relevant National or International standards for example ASTM E1, see Table 2.2. The thermometers should be calibrated to 0.005 °C by the National Metrology Institute maintaining the appropriate national standards for temperature. However, calibrated platinum resistance thermometer along with Resistance Bridge is preferable.

Other calibrated thermometers at 46 °C or at 121 °C as required for standardising oil viscosities may also be used. The thermometers used in viscosity measurement may conform to ASTM standards as mentioned in Table 2.2.

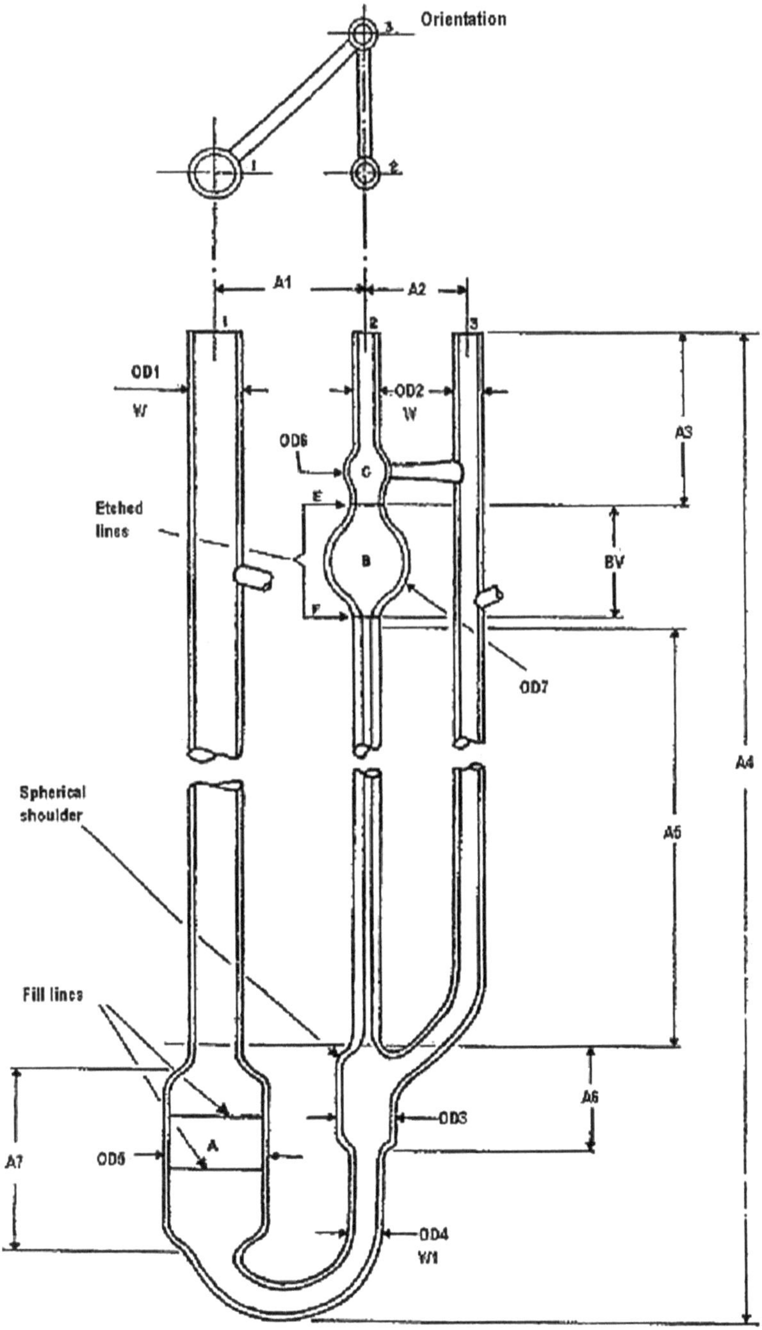

Fig. 2.4 Ubbelohde master viscometer

Table 2.2 ASTM standards for various thermometers

Thermometer no.	Test temperature (°C)	Thermometer no. (°C)	Test temperature (°C)
Above 0 °C		Below 0 °C	
ASTM132C	150	ASTM128C	0
ASTM110C	135	ASTM72C	−17.8
ASTM122C	100	ASTM127C	−20
ASTM129C	93.3	ASTM126C	−26.1
ASTM48C	82.2	ASTM73C	−40
ASTM47C	60	ASTM74C	−53.9
ASTM29C	54.5		
ASTM46	50		
ASTM120C	40		
ASTM28C	37.8		
ASTM118C	30		
ASTM45C	25		
ASTM44C	20		

2.4.3 Bath

A thermo-controlled bath, of minimum depth 650 mm (deep enough to immerse the master viscometer so that upper fiducial mark is at least 50 mm below the free water or oil surface) and wide enough to accommodate three to four viscometers, with ports for thermometers, is suitable. Proper viscometer stands should be provided such that a master viscometer is firmly held in reproducible position and remain vertical within ±15 min of arc. The bath may contain water or other suitable transparent oil as a thermo-static liquid.

The efficiency of stirring and the balancing between the heat loss and input must be such that the temperature of the working liquid does not vary by more than ±0.01 °C over the length of the viscometer or from one viscometer position to another. The working section of the bath should be shielded from the direct radiation from heaters and light. A standard platinum resistance thermometer of 450 mm nominal length may be used to ensure that the variation in temperature does not exceed ±0.01 °C.

2.4.4 Timer

Either a spring-wound or electric timer with measuring range of 300–10,000 s with an accuracy of ±0.03 % is used. The stop watch should be calibrated and used when in fully but not tightly wound position. The electric timers must be operated on frequency controlled circuits. Commercial power sources whose frequencies are intermittently and not continuously controlled are not satisfactory.

2.4.5 Cleaning Agents

1. Acetone used for drying should be of reagent grade.
2. Chromic acid used for cleaning should be freshly prepared. In 35 ml of saturated solution of technical grade sodium dichromate, 1 litre of concentrated H_2SO_4 (sulphuric acid) of relative density 1.84 is poured carefully. To avoid disposal problem of solutions containing chromium, any strong oxidising agent containing no chromium may be used.
3. Petroleum spirit or other solvent completely miscible with oils should be of reagent grades.

2.4.6 Standard Liquids

2.4.6.1 Primary Standard

Freshly prepared on the day of use, deionised or distilled water and passed through micro filters, is taken as primary standard. Its kinematic viscosity is assumed as 1.0034 mm^2/s. The water should conform to ISO 3606: 1987 "Water for analytical laboratory use- specifications and test methods".

2.4.6.2 Standard Oils

Chemically stable oils are selected to cover the required viscosity range. These should be transparent and have vapour pressure <10 mm of Hg at 40 °C. The oils are stored away from the light preferably in glass bottles.

2.5 Detailed Procedure

2.5.1 Calibration of Master Viscometers

The master viscometer having viscometer constant in the range of 0.001 –0.003 mm^2/s^2 is chosen and its viscometer constant is determined with the help of de-ionised distilled water at 20 °C.

2.5.1.1 Preparing the Bath

The water bath is maintained at a constant temperature, such that after applying proper necessary corrections to thermometers, the temperature of the bath is 20 ± 0.01 °C. Two thermometers should be used to check any temperature variation.

2.5.1.2 Cleaning of Viscometer

Master viscometer is first cleaned with chromic acid and then with cleaning solution to remove organic deposits. After that the viscometer is thoroughly rinsed with distilled water and acetone and finally dried with a stream of filtered air. Similarly clean a 50 ml flask and rinse it thoroughly with distilled water. Add freshly distilled water to be used as primary standard and heat it almost to its boiling point to remove any dissolved gasses. Cover the flask to prevent entrance of dust and allow it to cool to 20 ± 3 °C for Cannon master viscometer.

2.5.1.3 Charging the Cannon Master Viscometer

Connect rubber tubing to the large arm J (Fig. 2.3). Invert the viscometer and dip the end of small arm into the flask of water. Apply suction to the tube J and draw water through the bulb B and D to the etched line E. Turn the viscometer in the upright position, wipe the tube A clean, insert the viscometer into its holder and place it in a constant temperature bath maintained at 20 ± 0.01 °C.

Support the viscometer firmly and align it in a vertical position with the aid of a plumb line. The length of the plumb line should be about 560 mm.

2.5.1.4 Charging the Ubbelohde Master Viscometer

Tilt the viscometer about 30° from the vertical with bulb A bellow the capillary (Fig. 2.4) and then introduce enough water into the large arm 1 to bring the level up to the lower filling mark. The level should not be above the upper filling mark when the viscometer is returned to the vertical position and the water has been drained from the tube 1. Charge the viscometer in such a manner that the U- tube at the bottom fills completely with out trapping any air.

Fix the viscometer into the holder and place the holder in the constant temperature bath. Align the tube 1 in a vertical position with a plum line as described for the Cannon viscometer.

2.5.1.5 Waiting for Thermal Equilibrium

Allow the charged viscometer to stand in the bath long enough for the sample to reach bath temperature. Fifteen minutes are normally sufficient to reach thermal equilibrium.

2.5.1.6 Measurement of Efflux Time

With gentle pressure or vacuum, force the water about 5 mm above the upper timing mark. Avoid splashing of water in the upper bulb or forming any bubbles in the water. When using the Ubbelohde viscometer, hold a finger over the upper end of tube 3 during this operation; then remove it and immediately place it over the tube 2 until water drops away from the lower end of the capillary. The efflux time T of water from the upper to lower timing mark is measured. The time of efflux should be more than 300 s and be measured to the nearest 0.1 s.

Repeat this measurement two more times and take the average of three observations if the lowest and highest efflux times agree within 0.1 %. If the efflux times do not agree within 0.1 %, repeat the procedure paying particular attention to cleaning the viscometer, filtering the sample avoiding contamination during filling and afterwards, check the temperature control and the timing device.

Clean the viscometer and dry it with filtered air. Reload the viscometer and measure the efflux time in triplicate. Take the average only, if the maximum difference between any two values is <0.1 %. Calculate the viscometer constant C_{20} at 20 °C by the following equation

$$C_{20} = 1.0034/T. \tag{2.4}$$

2.5.2 *Calibration of Second Viscometer with Water*

Repeat the operations described in Sects. 2.5.1.1–2.5.1.6 with the second viscometer having a viscometer constant not >0.003 mm^2/s^2. This way we will have two master viscometers of similar viscometer constants.

2.5.3 *Determination of Viscosity of Oil (Measurement at 40 °C)*

The step consists of the determination of viscosity of a suitable oil standard at 40 °C.

2.5.3.1 Preparation of Oil Standard

Choose the oil having kinematic viscosity of about 3 mm^2/s at 40 °C. Filter a portion into a clean beaker through a 200-mesh (75 μm) sieve, or through other suitable filter to remove dirt and sediment. If necessary adjust the temperature of the oil to within 3 °C of 40 °C.

2.5.3.2 Preparation of Bath

The water bath is maintained at a constant temperature, such that after applying proper necessary corrections to thermometers, the temperature of the bath is at 40 ± 0.01 °C. Two thermometers should be used to check any temperature variation.

The bath temperature is to be maintained at 40 ± 0.01 °C, as the kinematic viscosity of petroleum oils are traditionally measured at this temperature and viscosity standard oils calibrated at 40 °C are commonly available. Master viscometers may, however, be calibrated at any temperature such as 100 °C for highly viscous oils or at below zero °C for very low viscosity oils.

2.5.3.3 Charging the Viscometer

Depending upon the type of viscometer, use either of the method described in Sects. 2.5.1.3 or 2.5.1.4 to charge the viscometer.

2.5.3.4 Waiting for Thermal Equilibrium

Allow the charged viscometer to stand in the bath long enough for the sample to reach bath temperature. Fifteen minutes are normally sufficient to reach thermal equilibrium.

2.5.3.5 Measurement of Efflux Time

The efflux time T of the oil from the upper to lower timing mark is measured. The time of efflux should be <1000 s but more than 300 s and be measured to the nearest 0.1 s.

Repeat this measurement two more times and average if the lowest and highest efflux times agree within 0.1 %. If the efflux times do not agree within 0.1 %, repeat the procedure paying particular attention to cleaning of the viscometer, filtering the sample avoiding contamination during filling and afterwards, checking the temperature control and the timing device.

Clean the viscometer with petroleum spirit followed by acetone and dries it with filtered air. Reload the viscometer and measure the efflux time in triplicate. Take average if the maximum difference between any two values is <0.1 %. Let the efflux time of the oil sample be T_1. Calculate the viscosity of the oil.

2.5.4 Measurement with Second Viscometer

Using second calibrated viscometer, repeat steps Sects. 2.5.3.1–2.5.3.4 and measure the efflux time of the same oil sample as in Sect. 2.5.3 and calculate the kinematic viscosity of the oil. Take the mean of the viscosity values obtained in Sects. 2.5.3 and 2.5.4.

Another oil of similar viscosity is taken and its kinematic viscosity is determined with the help of both viscometers. These two oils are used to determine the viscometer constants of another two similar viscometers of the next step.

2.5.5 *Corrections and Calculation of Kinematic Viscosity at 40 °C*

2.5.5.1 Different Temperatures of Measurement and Statement

The viscosity of all liquids is temperature dependent. Hence viscosity of a liquid is stated at a given standard temperature. The bath is maintained at the standard temperature however the actual bath temperature may still differ from the standard temperature. Hence a correction is applicable due to this cause. The coefficient of viscosity is given by

$$\alpha_v = (\nu_{40} - \nu_{37.78})/2.22\nu_{40} \tag{2.5}$$

α_v is the coefficient of increase in viscosity per kelvin, ν_{40} and $\nu_{37.78}$ are viscosity values of the liquid under consideration at 40 °C and 37.78 °C respectively. Typical values of viscosity coefficients of some standard oils is given in column 4 of Table 2.3.

2.5.5.2 Buoyancy Correction

The buoyancy correction is applicable to both Cannon and Ubbelohde viscometers. In kinematic viscometers the driving head of the liquid is slightly reduced by the counter-poised head of air in the empty arm of the viscometer. The buoyancy effect is measured by the density of air divided by the density of the liquid. In going from water at 20 °C to oil at 40 °C, the relative buoyancy correction c_b is

$$c_b = \left(\frac{\sigma_{20}}{\rho_{w20}} - \frac{\sigma_t}{\rho_{ot}}\right) \tag{2.6}$$

Here σ_{20} and σ_t density values of air at 20 °C and t °C respectively
ρ_{w20} density of water at 20 °C,
ρ_{ot} density of oil at t °C and
c_b is relative buoyancy correction as a fraction of the driving head.

In general relative buoyancy correction for a viscometer calibrated with one oil at one particular temperature but is used for another liquid of different density and at different temperature is given by

Table 2.3 Coefficient of viscosity and relative buoyancy corrections

Viscosity standard	Kinematic viscosity		Coeff. of viscosity	Density g/ml		Buoyancy correction
	37.78 °C	40 °C		20 °C	40 °C	
S3(a)	3.0	2.9	0.0155	0.866	0.854	–
N10	11	10	0.045	0.884	0.871	0.000 03
N35	35	32	0.0422	0.868	0.854	−0.000 03
N100	110	97	0.0604	0.882	0.870	0.000 02
N350	340	310	0.0436	0.843	0.831	−0.000 06
N1000	550	940	0.0415	0.850	0.838	0.000 01
N2500	1500	2500	0.0400	0.849	0.837	−0.000 00
S8000	8000	6700	0.0873	0.890	0.879	0.000 06
N10200	4900	10200	0.1082	0.891	0.880	0.000 00
S30000	28000	23000	0.0979	–	0.886	0.000 01

$$c_b = \left(\frac{\sigma_{t1}}{\rho_{t1}} - \frac{\sigma_{t2}}{\rho_{t2}} \right). \tag{2.7}$$

Example:
Buoyancy correction from water at 20 °C to oil at 40 °C
We know that
Density of water at 20 °C = 998.4081 at 40 °C = 992.2166 kg/m^3
Density of air at 20 °C = 1.199217 kg/m^3 at 40 °C = 1.111770 kg/m^3
Relative Buoyancy correction to viscometer calibrated by water at 20 °C but used to determine the viscosity of oil at 40 °C = 1.199217/998.41−1.111770/748
= 0.00120 112 − 0.00148632
= −0.000 285 or 0.03 %

A particular standard oil of certain density is used to determine the viscometer constant of the given viscometer and the same viscometer is used to determine viscosity of another standard oil of different density. Hence relative buoyancy correction due to difference in densities of two oils becomes applicable.

The coefficients of viscosity and relative buoyancy corrections of some standard oils are given in Table 2.3. Column 4 gives the coefficients of viscosity and buoyancy corrections are given in column 7.

Here we see that for most petroleum oils the relative buoyancy correction varies from 0.006 to 0.001 %.

2.5.5.3 Temperature Correction

There are two situations when temperature correction is to be applied. (1) some viscometers are charged (filled) at room temperature say around 25 °C but are used at 20 or 40 °C, and (2) a viscometer is calibrated at one temperature but is used at different temperature.

The effect of thermal expansion of glass capillary viscometer is self compensatory and usually may be neglected. However, for the viscometers for which the working volume is fixed at 20 ± 3 °C but which are used at another temperature, the driving head changes with expansion or contraction of the liquid. The correction c_t is given as follows:

$$c_t = \frac{V(\rho_t - \rho_{20})}{\pi R^2 h \rho_t} \tag{2.8}$$

where:

c_t is the temperature correction expressed as a fraction of the driving head
V total volume of the fill at 20 °C in cm^3
ρ_{20} is density of the liquid filled at 20 °C
ρ_t density of the liquid at test temperature
R inside radius of the reservoir in cm and
h driving head in cm at 20 °C

For dimensions of the Cannon master viscometers, the above equation reduces to

$$c_t = 0.023 \frac{(\rho_t - \rho_{20})}{\rho_t} \tag{2.9}$$

Temperature correction for a viscometer calibrated with water at 20 °C and used with oil at 40 °C is –0.0004.

Temperature correction for a viscometer charged at room temperature 23.5 °C with water but calibrated at 20 °C is 0.000 02, which is negligible. However for high viscosity liquids, the liquid is filled at room temperature but measurements are taken at 100 °C. In that case correction is substantial. Further in such cases, the coefficient of expansion of viscometer glass is also taken into account. The modified formula in this case is:

$$c_t = \frac{V(\rho_t - \rho_{20}) - 3\alpha(t_f - t_m)}{\pi R^2 h \rho_t} \tag{2.10}$$

Here t_f and t_m are filling and measurement temperatures. The α is coefficient of expansion of glass.

Temperature correction for a viscometer charged at room temperature 20 °C with a oil but calibrated at 40 °C is 0.000 3, which is not negligible.

Data for the above example

$\rho_{w20} = 0.998\ 201\ g/cm^3$, $\rho_{w23.5} = 0.997.446\ g/cm^3$
$\rho_{oil20} = 0.884\ g/cm^3$ and $\rho_{oil40} = 0.871\ g/cm^3$.

2.5.5.4 Surface Tension Correction

The effective driving head will change with surface tension, if the radii of the capillary are not equal at E and F timing marks or at F and reservoir. The effect is a maximum with water, as water has higher surface tension than most of the petroleum oils. One such correction is given in Chap. 1 and is reproduced here.

$$c_s = \frac{2}{gh}\left(\frac{1}{r_u} - \frac{1}{r_L}\right)\left(\frac{\Gamma_c}{\rho_c} - \frac{\Gamma_t}{\rho_t}\right) \tag{2.11}$$

where:

c_s is surface tension correction expressed as the fraction of the driving head
r_u and r_L are the radii at the upper and lower timing marks respectively
Γ_c is surface tension of calibrating liquid (in this case, it is water)
Γ_t is the surface tension of oil used at 40 °C (or of under test liquid)
ρ_c and ρ_t are respective density of calibrating and under-test oil
h is driving head g is local acceleration due to gravity.

In cannon master viscometer, the approximate surface tension correction for most of petroleum oils is 0.0014 when correcting from water at 20 °C to oil at 40 °C. The surface tension correction is greatest when the viscometer was calibrated at 20 °C with water and used for determining the viscosity of oil at 40 °C.

In Ubbelohde viscometers, the surface tension correction will be nearly zero if the diameter of the meniscus in the vent bulb below the capillary is not <75 % that of the meniscus in the efflux bulb at the maximum diameter. If the diameter of the lower meniscus varies from 70 to 80 % of the maximum upper diameter, the surface correction c_s will vary from −0.0002 to + 0.0002; when the viscometer is made as in the Fig. 2.4. The surface tension correction c_s, should be within these limits. If the diameter of the lower meniscus varies from 60 to 90 %, the surface tension correction c_s will vary from −0.0005 to +0.0005.

2.5.6 Calculation of Kinematic Viscosity

2.5.6.1 Cannon Master Viscometer

For a canon master viscometer calibrated with water, the kinematic viscosity of oil is obtained from the following equation:

$$v = C_c(1 + c_b + c_t + c_s) \cdot T \tag{2.12}$$

2.5.6.2 Ubbelohde Master Viscometer

For Ubbelohde master viscometer calibrated with water, kinematic viscosity is given as:

$$v = C_c(1 + c_b) \cdot T \tag{2.13}$$

Here C_c is the viscometer constant at 20 °C of the master viscometer calibrated with water

c_b is the relative buoyancy correction to C_c
c_t is the relative temperature correction to C_c
c_s is the relative surface tension correction to C_c

For the Cannon master viscometer, the value of $(1 + c_b + c_t + c_s)$ will be around 1.0009 instead of 1 i.e. 0.09 % will be the correction. The total correction is of the same order as the repeatability observed for determination of efflux time.

Similarly for Ubbelohde master viscometer the correction factor is 0.9998 instead of 1. Here the total correction is much less than the repeatability for the determination of efflux time.

With unbiased procedure, the values of kinematic viscosity of the oil standard determined by two master viscometers should be within 0.1 %.

Some typical values of these corrections given by Fujita [4] are tabulated in Table 2.4.

2.6 Standards Maintained at NPLI

The National Physical laboratory of India (NPLI) maintains a number of viscometers and standard oils. The viscometer constant of each viscometer and viscosity of standard oil is determined by the procedure described in Sect. 2.4 above.

2.6.1 Viscometers

The viscometers presently maintained at NPLI with their respective viscometer constants are listed in Table 2.5.

2.6.2 Standard Oils

In addition to the aforesaid viscometers, the National Physical Laboratory of India maintains standard oils of stable viscosity and density of various companies dealing with viscometers and standard oils. The viscosity of the oil is determined

Table 2.4 Relative corrections to viscometer constant due to various parameters

Oil to oil at −40 °C	Oil to oil at 30 °C	Oil to oil at 100 °C
$c_b = 0.000\ 28$	$c_b = 0.000\ 02$	$c_b = 0.000\ 32$
$c_f = -0.001\ 91$	$c_t = 0.000\ 14$	$c_t = 0.001\ 88$
$c_e = -0.000\ 19$	$c_e = 0.000\ 00$	$c_e = 0.000\ 26$
$c_s = 0.000\ 00$	$c_s = 0.000\ 00$	$c_s = 0.000\ 00$

Table 2.5 Standard viscometers

S no.	Number	Viscometer constant (mm^2/s^2)	Uncertainty (%)
1	O_{C2}	0.002856	0.035
2	O_{C3}	0.003060	0.035
3	I_2	0.009749	0.035
4	I_{C1}	0.029008	0.035
5	I_{C2}	0.029541	0.035
6	II_1	0.101247	0.035
7	II_2	0.101214	0.035
8	II_{C1}	0.318355	0.035
9	II_{C2}	0.307737	0.035
10	III_1	0.974006	0.035
11	III_2	0.980541	0.035
12	III_{C1}	3.120440	0.035
13	III_{C3}	2.989015	0.035
14	IV_1	9.665991	0.035
15	IV_{a3}	Unknown	0.13×10^{-6}

and cross checked by using appropriate viscometers. The standard oils available at NPL India with their approximate viscosity and density values are given in Table 2.6.

The correction is necessary for Cannon type viscometer, when the test temperature (say 40 °C) differs from the calibration temperature (20 °C).

2.7 Propagation of Uncertainty in Establishing the Viscosity Scale

2.7.1 Expression for Uncertainty in the nth Step

To establish the viscosity scale, we start with water as primary standard of known viscosity and determine the viscometer constant of a suitable master viscometer MV_1 by observing the efflux time at 20 °C by using the following equation:

Table 2.6 Standard oils of known viscosity and density available at NPL India

Viscosity standards	Kinematic viscosity (mm²/s)		Density (g/ml)	
	20 °C	40 °C	20 °C	40 °C
S3(a)	4.6	2.9	0.868	0.854
S6	11	5.7	0.878	0.864
N10	21	10	0.884	0.871
N14	30	14	0.812	0.800
N35	87	32	0.868	0.854
S60	160	54	0.876	0.863
N100	330	97	0.882	0.870
S200	550	180	0.840	0.827
N250	770	250	0.842	0.830
N350	1000	310	0.843	0.831
S600(c)	1700	520	0.847	0.835
N1000	3400	940	0.850	0.838
S2000	8300	1600	0.877	0.865
N4000	20000	3400	0.883	0.872
S8000	41000	6700	0.890	0.879
N15000	77000	13000	0.894	0.823
S30000	–	23000	–	0.886

$$v = C_1 T_1 - \frac{E}{T_1^2} \tag{2.14}$$

Second term on the right hand side of (2.14) is the kinetic energy correction, which is small in comparison to the first term. So this term may be neglected for the purpose of estimating uncertainty. Hence neglecting the kinetic energy term in (2.14), we get

$$C_1 = \frac{v}{T_1} \tag{2.15}$$

Giving

$$\log C_1 = \log v - \log T_1 \tag{2.16}$$

On differentiating we get

$$\frac{\delta C_1}{C_1} = \frac{\delta v}{v} - \frac{\delta T}{T_1} \tag{2.17}$$

By applying the law of variances and replacing relative variances by the square of the relative standard uncertainties, we get

$$U_{C1}^2 = U_v^2 + U_{T1}^2 \tag{2.18}$$

Here U_{C1}, U_v and U_{T1} are the relative standard uncertainties of C_1, v and T_1 respectively.

The equation (2.15) is true if kinetic energy correction is negligible. Even in case of master viscometers of fairly large length and efflux time, the kinetic energy correction is about 0.004–0.006 % of the first term on right hand side of (2.15) [4]. This value appears to be small but not negligible especially when the relative uncertainty of the viscometer constant of the master viscometer calibrated with water is <0.01 %. Therefore this correction is applied in the first step and is neglected in the second and subsequent steps as its contribution in the determination of viscometer constant of the viscometer or viscosity value of the standard liquid is 0.01 % or less.

The calibrated viscometer MV_1 is used to determine the viscosity of the first standard liquid RL1 at 40 °C. The equation (2.19A) is used if the master viscometer is Cannon type and (2.19B) is used if the master viscometer is Ubbelohde type.

$$\nu_{RL1} = C_1(1 + c_b + c_t + c_s) \cdot T_2 = C_{\text{mod}} \cdot T_2 \tag{2.19A}$$

$$\nu_{RL1} = C_1(1 + c_b) \cdot T_2 \tag{2.19B}$$

where

$$C_{\text{mod}} = C_1(1 + c_b + c_t + c_s), \tag{2.20}$$

$$c_b = \frac{\sigma_{20}}{\rho_{w20}} - \frac{\sigma_t}{\rho_{Lt}} \tag{2.21}$$

is the buoyancy correction.

$$c_t = \frac{V_f\{(\rho_f/\rho_m - 1) - 3\alpha(t_m - t_f)\}}{\pi R^2 h}$$
$$c_t = \frac{V_f \Delta h}{Vh}\left\{\frac{\rho_f - \rho_m}{\rho_m} - 3\alpha(t_m - t_f\right\} \tag{2.22}$$

is the temperature correction due to different measurement and liquid filling temperatures.

$$c_s = \frac{2}{gh}\left(\frac{1}{r_u} - \frac{1}{r_L}\right)\left(\frac{T_{w20}}{\rho_{w20}} - \frac{T_t}{\rho_t}\right) \tag{2.23}$$

is the surface tension correction.

Here:

V_f is total volume of the liquid filled and V is the volume of the timing bulb;

h is the mean effective head of the viscometer and Δh is change in effective head by the volume V;

σ_{20} and σ_t are density of air at 20 °C and temperature t °C respectively;

ρ_{20} is the density of water at 20 °C and ρ_t is density of the first standard liquid at temperature t °C;

R is inside radius of the reservoir; r_u and r_L are the radii of the capillary at the upper and lower graduation marks respectively;

T_c is the surface tension of water and T_t is the surface tension of the first standard liquid.

From (2.20) and following the law of variances, we get

$$U_{\mathrm{mod}}^2 = U_{C1}^2 + U_{cb}^2 + U_{ct}^2 + U_{cs}^2 \tag{2.24}$$

U_{cb}, U_{ct} and U_{cs} is the standard uncertainty due to buoyancy, change in temperature and change in surface tension respectively.

From (2.19A or 2.19B) the relative standard uncertainty of the first standard liquid U_{RL1} is given as

$$U_{RL1}^2 = U_{\mathrm{mod}}^2 + U_{T2}^2 \tag{2.25}$$

Using (2.18) and (2.24) the equation (2.25) becomes

$$U_{RL1}^2 = (U_v^2 + U_{corr}^2) + (U_{T1}^2 + U_{T2}^2) \tag{2.26}$$

Here

$$U_{corr}^2 = U_{cb}^2 + U_{ct}^2 + U_{cs}^2 \tag{2.27}$$

The process is repeated with subsequent viscometer and second standard liquid giving the viscometer constant C_2 and its uncertainty is

$$U_{C2}^2 = U_{RL1}^2 + U_{T3}^2 \tag{2.28}$$

or

$$U_{C2}^2 = \left(U_v^2 + U_{corr}^2\right) + \left(U_{T1}^2 + U_{T2}^2 + U_{T3}^2\right) \tag{2.29}$$

The viscosity of second standard oil v_{RL2} is determined and its uncertainty will be given as

$$U_{RL2}^2 = U_{C2}^2 + U_{T4}^2 + U_{corr}^2 \tag{2.30}$$

Additional term U_{corr}^2 appeared in (2.27) as corrections given in (2.21), (2.22) and (2.23) have not been applied to C_2 and their contribution is taken as the uncertainty in C_2.

The equation (2.30) may be simplified as

$$U_{RL2}^2 = \left(U_v^2 + 2U_{corr}^2\right) + \left(U_{T1}^2 + U_{T2}^2 + U_{T3}^2 + U_{T4}^2\right) \tag{2.31}$$

So in the nth step for nth viscometer and nth standard liquid, uncertainties will be related as

$$U_{Cn}^2 = U_{RLn-1}^2 + U_{T(2n-1)}^2 \tag{2.32}$$

or

$$U_{Cn}^2 = \left\{U_v^2 + (n-1)U_{corr}^2\right\} + \sum_{r=1}^{r=2n-1} U_{Tr}^2 \tag{2.33}$$

$$U_{RLn}^2 = U_{Cn}^2 + U_{corr}^2 + U_{T2n}^2 \tag{2.34}$$

or

$$U_{RLn}^2 = \left(U_v^2 + nU_{corr}^2\right) + \sum_{r=1}^{r=2n} U_{Tr}^2 \tag{2.35}$$

Assuming that we are able to maintain same relative standard uncertainty in measurement of efflux time through out the step up process i.e. all U_{T2n} are equal for each value of n i.e. $U_{T1} = U_{T2} = U_{T3} = \cdots\cdots = U_{Tn} = U_T$, then (2.33) and (2.35) respectively become

$$U_{Cn}^2 = \left\{U_v^2 + (n-1)U_{corr}^2\right\} + (2n-1)\cdot U_T^2 \tag{2.36}$$

$$U_{RLn}^2 = \left(U_v^2 + nU_{corr}^2\right) + 2n\cdot U_T^2. \tag{2.37}$$

2.7.2 Planning for Uncertainty

There are two methods of measuring the given measurand and evaluation of uncertainty. First is that you measure with the supplied equipment and calculate the measurement uncertainty. Second is that you fix the acceptance tolerance of the observations and choose the proper equipment to attain it. In viscosity measurement, the measurand is efflux time. The ASTM D2162-06 [2] adopts the second method. The acceptance tolerance in the measurement of efflux time is taken as 0.1 %. That is no two efflux times differ by more than 0.1 %. This means that any legitimate efflux time should lie with equal probability any where within the range ±0.05 %, i.e. the efflux time follows a rectangular distribution with $0.0005{\cdot}T$ as the semi-range.

Relative standard deviation in measurement of efflux time, therefore is

$$U_T = 0.0005\big/\sqrt{3} = 0.00029 \tag{2.38}$$

We may remember that only first viscometer is used at two different temperatures with two liquids, one is water and the other is oil. Surface tension and density of any oil differ very much from that of water. Hence the applicable correction to the viscometer constant of first viscometer is a maximum. All other viscometers are used at the same temperature 40 °C and with oils of similar surface tensions and densities. Hence applicable corrections will be much smaller than those applied on the first viscometer.

The values of different corrections relative to C_1 from water to the oil for a typical case discussed in [4] are as follows:

$$\begin{aligned}\text{Buoyancy correction } c_b &= 0.000\,32\\ \text{Thermal expansion of liquid } c_t &= -0.000\,14\\ \text{Surface tension correction } c_s &= 0.001\,52\end{aligned} \tag{2.39}$$

These corrections are applied to the first viscometer which was used for determining the kinematic viscosity of first oil standard. For subsequent steps the corrections are not applied but their contribution is seen as uncertainty U_{corr}. Taking that all corrections will lie with equal probability within the semi range equal to the applicable correction for the first viscometer, we get the following components of uncertainty.

Uncertainty due to buoyancy correction is given by

$$U_{cb} = \frac{0.00032}{\sqrt{3}} = 0.00018 \tag{2.40}$$

Uncertainty due to temperature is given by

$$U_{ct} = \frac{0.00014}{\sqrt{3}} = 0.00008 \tag{2.41}$$

Uncertainty due to change in surface tensions of the liquids used is given by

$$U_{cs} = \frac{0.00152}{\sqrt{3}} = 0.00088 \tag{2.42}$$

Combining the two uncertainties we get the relative standard combined uncertainty as

$$\begin{aligned}U_{corr} &= \sqrt{(0.00018)^2 + (0.00088)^2}\\ &= 0.00090\end{aligned} \tag{2.43}$$

Substituting the values of U_{corr} and U_T in (2.22) we get the uncertainty in viscometer constant in the nth step as

$$U_{Cn} = 10^{-5} \cdot \sqrt{9782n - 8941} \tag{2.44}$$

For viscometer constant in the 9th step, the relative standard uncertainty is given as

$$U_{Cn} = 0.281\% \tag{2.45}$$

Similarly the relative standard uncertainty of the kinematic viscosity of the standard liquid used in the nth step is given as

$$U_{RLn} = 10^{-5} \cdot \sqrt{9782n} \tag{2.46}$$

For viscosity of 9th standard oil, the relative standard uncertainty is given as

$$U_{RLn} = 0.296\ \% \tag{2.47}$$

2.7.3 Correction Due to Different Measuring and Stated Temperatures

We know that the viscosity of a liquid depends upon its temperature. Hence in addition of the aforesaid corrections there is one more source of correction due to lack of ability of adjusting the temperature of the bath exactly at the given standard temperature t_s. For example when we are assuming that bath has been adjusted at $t_s = 40$ °C its actual temperature may be $t_m = 39.812$ °C. The viscosity of the oil standard determined will then be at 39.812 °C and not at 40 °C. If α_v be the coefficient of increase in viscosity for that oil then the correction C_v to be applied is given as

$$C_v = \alpha_v(t_m - t_s) \tag{2.48}$$

This correction should be applied to the determined value of kinematic viscosity of the standard liquid at each step.

Temperature coefficient of viscosity [4]

for water	at 20 °C	0.0243 K^{-1}
for JS 160000	at 40 °C	0.098 K^{-1}
for JS 2.5	at 20 °C	0.020 K^{-1}
for JS 52000	at 30 °C	0.092 K^{-1}
for S 600	at 100 °C	0.032 K^{-1}
for S3	at − 40 °C	0.081 K^{-1}

Maximum Correction for 0.01 °C difference in measurement and stated temperatures is

$$0.098(40.01 - 40.0) = 0.000\,98 \text{ or } 0.098\,\%$$
$$0.098\,\%$$

Minimum Correction for 0.01 °C difference in measurement and stated temperatures is

$$0.032\,\%.$$

2.7.4 Uncertainty in the Value of Viscosity of Water

In the above paragraphs the uncertainty in the value of the kinematic viscosity of water U_v has been taken as zero. The kinematic viscosity of water prior to 1990 was taken as 1.0038 mm^2/s [4] and now as per ISO/TR 3666, it is taken as 1.0034 mm^2/s [1], which is basically due to change in temperature scales. The uncertainty in measurement of viscosity of water as discussed in Chap. 8 is around 0.3 %.

This component of uncertainty may not be necessary if there is an agreement at the international level about a particular value of kinematic viscosity of water.

References

1. ISO (1998) Viscosity of water ISO/TR 3666. International organisation for Standardisation, Geneva
2. ASTM (2006) Standard Practice for Basic Calibration of Master Viscometers and Viscosity Oil Standards ASTM D2162-06
3. Gupta SV, Nangia SN, Sharma DC (1991) Establishment of viscosity scale. MAPAN-J Metrol Soc India 6:29–36
4. Fujita Y, Kurano Y, Fujii K (2009) Evaluation of uncertainty in viscosity measurements by capillary viscometers. Metrologia 46:237–248

Chapter 3
Capillary Viscometers

Abstract Glass capillary viscometers may be categorised in three groups. These are: (1) Oswald type in its various modified forms for transparent liquids; (2) Suspended level type in which capillary is always full, which ensures a uniform pressure head independent of the volume of the liquid charged into the viscometer and (3) Reverse flow type in which liquid moves from capillary to measuring bulb and therefore suitable for use for opaque liquids. Each category has number of sets of viscometers named after the names of inventors. Modified Ostwald and Suspended Level categories, each has seven types of viscometers covering a range 0.2–20,000 mm^2/s and 0.3–100,000 mm^2/s respectively. Reverse flow type have 4 sets of viscometers. The covering range goes up to 300,000 m^2/s. Importance of capillary viscometers is a very good repeatability. Hence are used to measure the viscosity of reference standard oils. Details of each type of viscometers have been discussed in this chapter.

3.1 Broad Classification

There are a large number of glass capillary viscometers. These can be broadly divided in three groups [1, 2]. Namely

A. Modified Ostwald Viscometers
B. Suspended Level Viscometers for transparent liquids
C. Reverse Flow Viscometers for transparent and Opaque liquids

Most of the viscometers are named after the name of their inventors.

S. V. Gupta, *Viscometry for Liquids*, Springer Series in Materials Science 194,
DOI: 10.1007/978-3-319-04858-1_3, © Springer International Publishing Switzerland 2014

3.2 Three Groups of Viscometers

3.2.1 Modified Ostwald Viscometers

These are used for determination of kinematic viscosity of transparent liquids. The group of Modified Ostwald Viscometers consist of

	Nomenclature	Range of Kinematic viscosity (mm^2/s)
1	Cannon–Fenske Routine Viscometer	0.5–20,000
2	Zeitfuchs Viscometer	0.6–3,000
3	SIL Viscometer	0.6–10,000
4	Cannon-Manning Semi micro Viscometer	0.4–20,000
5	U-Tube Viscometer	0.9–10,000
6	Miniature Viscometers	0.2–100
7	Pinkevitch Viscometer	0.6–17,000

3.2.2 Suspended Level Viscometers

Basic feature of suspended level viscometer is that the liquid under-test is suspended in the capillary and keeps it completely filled. This ensures constant driving head of liquid independent of the volume of the sample with which the viscometer is charged. The group of suspended level viscometers consist of

	Nomenclature	Range of Kinematic viscosity (mm^2/s)
1	Ubbelohde Viscometer	0.3–100,000
2	Cannon-Ubbelohde	0.5–100000
3	BS/IP/SL (S) Viscometer	1.05–10,000
4	BS/IP/MSL Viscometer	0.6–3,000
5	FitzSimons Viscometer	0.6–1,200
6	Atlantic Viscometer	0.7–5,000
7	Cannon-Ubbelohde Semi-micro Viscometer	0.4–20,000

3.2.3 Reverse Flow Viscometers

The liquid under-test flows into the timing bulb which was kept dry and not previously wetted by the liquid sample. This facilitate to measure the efflux time of the opaque liquid more accurately. In Ostwald viscometers the liquid flow from timing bulb through capillary and its thin opaque films make it difficult to measure time correctly.

The group of reverse flow viscometers for transparent and opaque liquids contains

	Nomenclature	Range of Kinematic viscosity (mm^2/s)
1	Zeitfuchs Cross-arm viscometer	0.6–100,000
2	Cannon-Fenske Viscometer	0.4–20,000
3	Lantz-Zeitfuchs Viscometer	60–100,000
4	BS/IP/RF U tube Reverse flow	0.6–300,000

Viscometer for opaque liquid only.

3.3 Modified Ostwald Viscometers

3.3.1 Cannon Fenske Routine Viscometers

A typical Cannon Fenske Routine viscometer is shown in Fig. 3.1. The set consists of 12 viscometers covering the range from 0.5 to 20,000 mm^2/s.

3.3.1.1 Dimensions

The labelled dimensions of each viscometer, as shown in Fig. 3.1, are the same and can be found out, for example, from ASTM D446. The internal dimensions vary. These are given in Table 3.1 [1].

3.3.1.2 Charging

To charge the Cannon-Fenske routine, invert the viscometer and apply suction to tube Y with tube X immersed in the liquid sample. Draw the sample to timing mark F. Mount the viscometer upright in the constant-temperature bath keeping tube Y vertical.

3.3.2 Zeitfuchs Viscometers

A typical Zeitfuchs viscometer is shown in Fig. 3.2.The set consists of 7 viscometers covering the range from 0.6 to 3,000 mm^2/s.

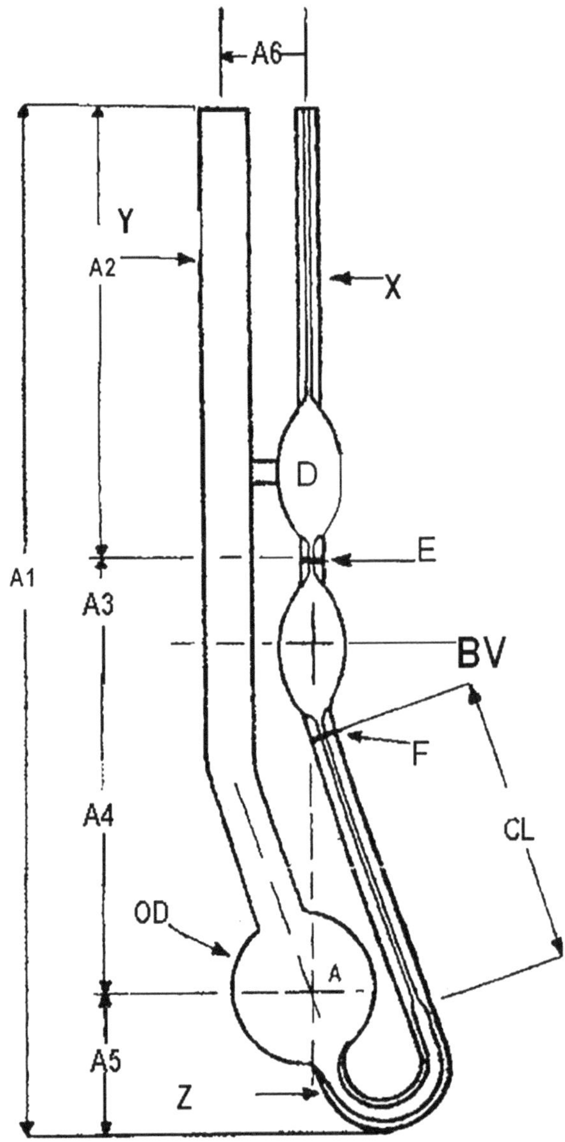

Fig. 3.1 Cannon-Fenske routine viscometer

3.3.2.1 Dimensions

The labelled dimensions of each viscometer, as shown in Fig. 3.2, are the same and can be found out, for example, from ASTM D446. The internal dimensions vary. These are given in Table 3.2 [1].

Table 3.1 Internal dimensions of ASTM viscometers

Size no.	Viscometer constant (mm^2/s^2)	Inside dia. of capillary CL (mm) (±2 %)	Inside dia. of tubes X, E, and Z (mm)	Bulb volume (mL) (±5 %)	
				D	BV
25	0.002	0.30	2.6–3.0	3.1	1.6
50	0.004	0.44	2.6–3.0	3.1	3.1
75	0.008	0.54	2.6–3.2	3.1	3.1
100	0.015	0.63	2.8–3.6	3.1	3.1
150	0.035	0.78	2.8–3.6	3.1	3.1
200	0.1	1.01	2.8–3.6	3.1	3.1
300	0.25	1.27	2.8–3.6	3.1	3.1
350	0.5	1.52	3.0–3.8	3.1	3.1
400	1.2	1.92	3.0–3.8	3.1	3.1
450	2.5	2.35	3.5–4.2	3.1	3.1
500	8	3.20	3.7–4.2	3.1	3.1
600	20	4.20	4.4–5.0	4.3	3.1

3.3.2.2 Charging

Mount the Zeitfuchs viscometer in the constant temperature bath, keeping tube Y vertical. Pour sample through tube Y to fill up to the mark G. Allow 15 min for the sample to attain bath temperature and to become free from air bubbles. Attach the vacuum line with stopcock and trap to tube M. Slowly draw the sample into timing bulb C by partially opening the stopcock in the vacuum line and partially closing tube X with the finger.

Allow the excess liquid to flow into bulb D and through tube M into the trap in the vacuum line. When the liquid in tube Y reaches a point 2–5 mm above filling mark J, hold it at this point by alternately closing the opening tube X to the atmosphere with the finger for some time, to permit the sample to drain from the walls of tube Y.

Adjust the working volume by drawing the meniscus at the bottom of the column of the liquid exactly to filling mark J, making sure that the sample completely fills the viscometer between mark J and the tip of the overflow in bulb D; after this final adjustment of the working volume, remove the finger and close or remove the connection to the vacuum source. The final adjustment may be more conveniently made by disconnecting the vacuum and applying pressure to the mounting tube Y by use of a rubber bulb.

3.3.3 SIL Viscometers

A typical SIL viscometer is shown in Fig. 3.3. The set consists of 8 viscometers covering the range from 0.6 to 10,000 mm^2/s.

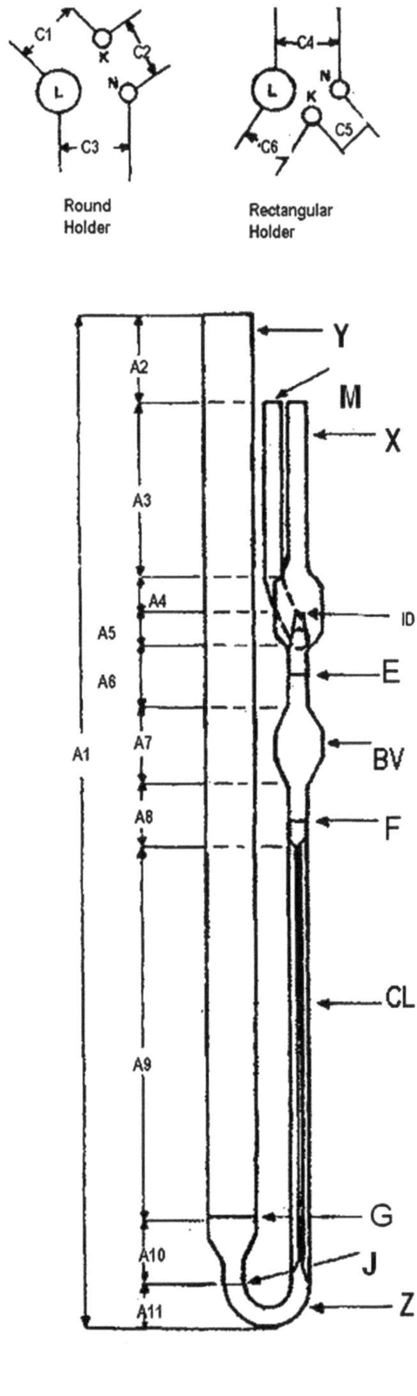

Fig. 3.2 Zeitfuchs Viscometer

Table 3.2 Internal dimensions of ASTM viscometers

Size no.	Viscometer constant (mm^2/s^2)	Inside dia. of working capillary CL (mm) (±2 %)	Inside dia. of tubes Z, E, and F (mm)	Volume of bulb BV (mL) (±5 %)
1	0.003	0.42	3.8–4.2	3.0
2	0.01	0.59	3.8–4.2	4.0
3	0.03	0.78	3.8–4.2	4.0
4	0.1	1.16	3.8–4.2	5.0
5	0.3	1.54	3.8–4.2	5.0
6	1.0	2.08	3.8–4.2	5.0
7	3.0	2.76	3.8–4.2	5.0

3.3.3.1 Dimensions

The labelled dimensions of each viscometer, as shown in Fig. 3.3, are the same and can be found out from the relevant standard, for example, ASTM D446. Only internal dimensions vary, these are given in Table 3.3 [1].

3.3.3.2 Charging

Charge the SIL viscometer by tilting it about 30° from the vertical, with bulb A below capillary CL. Introduce enough of the sample into tube Y for bulb A to fill completely and overflow into the gallery. Return the viscometer to the vertical position and mount it in the constant-temperature bath so that tube Y is vertical. The quantity of sample charged should be such that the level in the lower reservoir is 3–14 mm above opening S. The sample will rise in capillary CL somewhat higher than opening S. After the temperature equilibrium has been reached, remove any excess sample from the gallery by suction applied to tube M.

3.3.4 Cannon-Manning Viscometers

A typical Cannon-Manning viscometer is shown in Fig. 3.4. The set consists of 12 viscometers covering the range from 0.4 to 20,000 mm^2/s.

3.3.4.1 Dimensions

The labelled dimensions of each viscometer, as shown in Fig. 3.4, are the same and can be found out from the relevant standard, for example, ASTM D446. Only internal dimensions vary, these are given in Table 3.4 [1].

Fig. 3.3 SIL viscometer

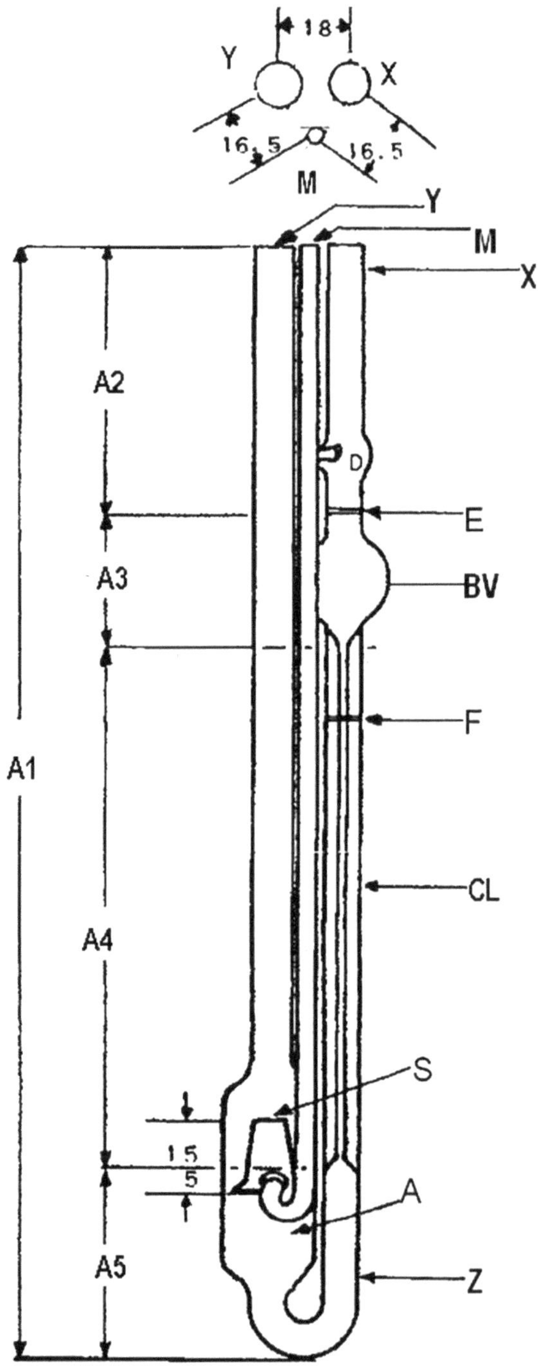

Table 3.3 Internal dimensions of ASTM viscometers

Size no.	Viscometer constant (mm^2/s^2)	Inside dia. of working capillary CL (mm) (±2 %)	Inside dia. of tubes E and Z (mm)	Volume of bulb BV (mL) (±5 %)
0C	0.003	0.41	4.5–5.5	3.0
1	0.01	0.61	4.5–5.5	4.0
1C	0.03	0.79	4.5–5.5	4.0
2	0.1	1.14	4.5–5.5	5.0
2C	0.3	1.50	4.5–5.5	5.0
3	1.0	2.03	4.5–5.5	5.0
3C	3.0	2.68	4.5–5.5	5.0
4	10	3.61	4.5–5.5	5.0

3.3.4.2 Charging

To charge a Cannon Manning viscometer, invert the viscometer and apply suction to tube Y with tube X immersed in the liquid sample. Draw the sample to the filling mark G. Mount the viscometer upright in the constant-temperature bath keeping tube Y vertical.

3.3.5 BS/U-Tube Viscometer

A typical BS/U-tube viscometer is shown in Fig. 3.5. The set consists of 8 viscometers covering the range from 0.9 to 10,000 mm^2/s.

3.3.5.1 Dimensions

The labelled dimensions of each viscometer, as shown in Fig. 3.5, are the same and can be found out from the relevant standard, for example, ASTM D446. Only internal dimensions vary, these are given in Table 3.5 [1].

3.3.5.2 Charging of BS/U

Mount the BS/U-tube viscometer in the constant-temperature bath keeping the tube Y vertical. Use a long pipette to minimize any wetting of tube Y above the filling mark G, fill bulb A with a slight excess of the sample. After allowing the sample to attain the bath temperature, adjust the volume of the sample to bring the liquid level within 0.2 mm of filling mark G by withdrawing the sample with a pipette.

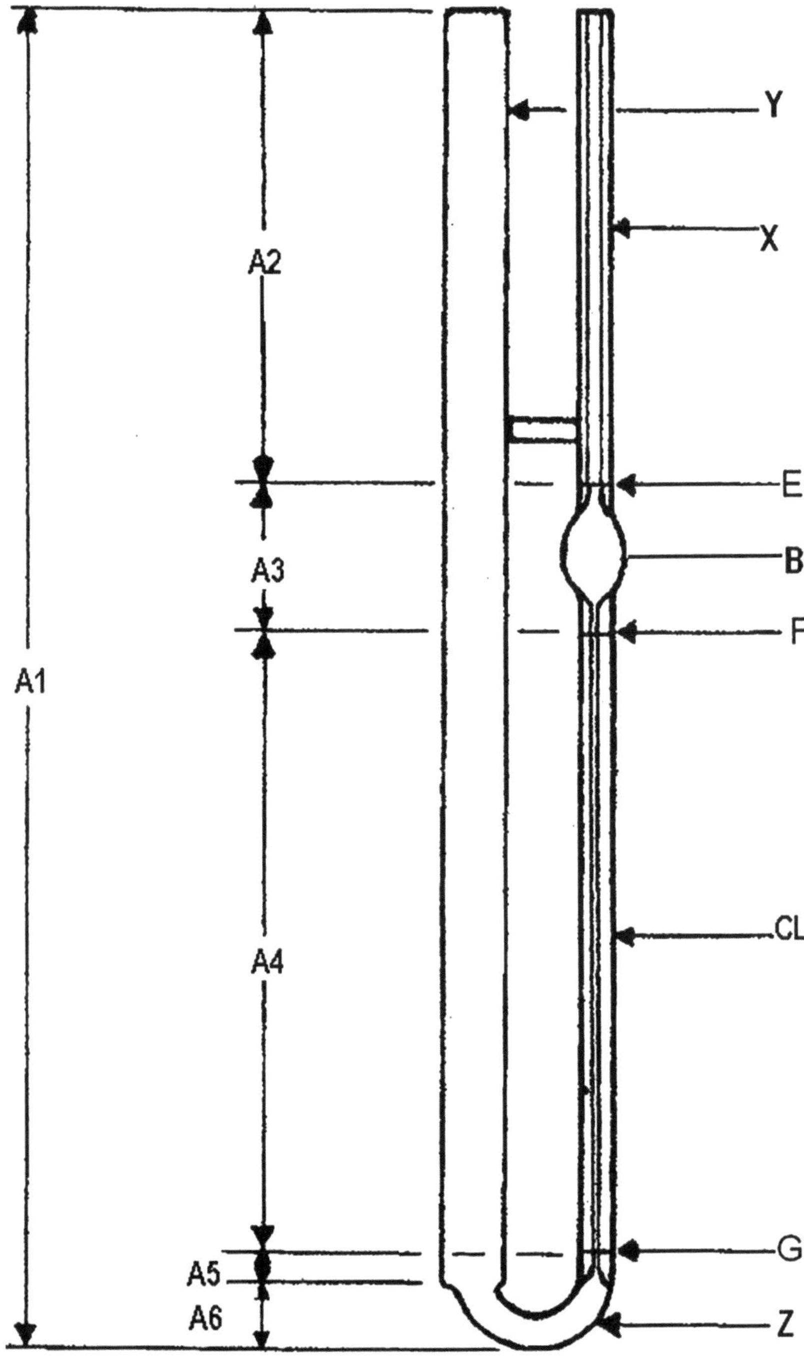

Fig. 3.4 Cannon-Manning viscometer

Table 3.4 Internal dimensions of ASTM viscometers

Size no.	Viscometer constant (mm^2/s^2)	Inside dia. of capillary CL (mm) (±2 %)	Inside dia. of tubes		Volume of bulb BV (mL) (± 5 %)
			X and F mm	Z mm	
25	0.002	0.22 ± 0.01	1.0–1.2	0.4–0.7	0.31
50	0.004	0.26 ± 0.01	1.0–1.2	0.5–0.8	0.31
75	0.008	0.31 ± 0.01	1.1–1.3	0.6–0.8	0.31
100	0.015	0.36 ± 0.02	1.2–1.4	0.7–0.9	0.31
150	0.035	0.47 ± 0.02	1.2–1.4	0.8–1.0	0.31
200	0.1	0.61 ± 0.02	1.4–1.7	0.9–1.2	0.31
300	0.25	0.76 ± 0.02	1.5–1.8	1.2–1.6	0.31
350	0.5	0.90 ± 0.02	1.8–2.2	1.5–1.8	0.31
400	1.2	1.13 ± 0.03	2.0–2.4	1.6–2.0	0.31
450	2.5	1.40 ± 0.04	2.2–2.6	2.0–2.5	0.31
500	8	1.85 ± 0.05	2.4–2.8	2.5–2.8	0.31
600	20	2.35 ± 0.05	3.0–3.4	2.7–3.0	0.31

3.3.6 Miniature Viscometers BS/U-Tube or BS/U/M

A typical BS/U-tube miniature viscometer is shown in Fig. 3.6. The set consists of 5 viscometers covering the range from 0.2 to 100 mm^2/s.

3.3.6.1 Dimensions

The labelled dimensions of each viscometer, as shown in Fig. 3.6, are the same and can be found out from the relevant standard, for example, ASTM D446. Only internal dimensions vary, these are given in Table 3.6 [1].

3.3.6.2 Charging

The BS/U/M miniature viscometer is mounted in the constant-temperature bath, keeping the tube Y vertical. Use a long pipette to minimize any wetting of tube Y above the filling mark G, fill bulb A with a slight excess of the sample. After allowing the sample to attain the bath temperature, adjust the volume of the sample to bring the liquid level within 0.2 mm of filling mark G by withdrawing the sample with a pipette.

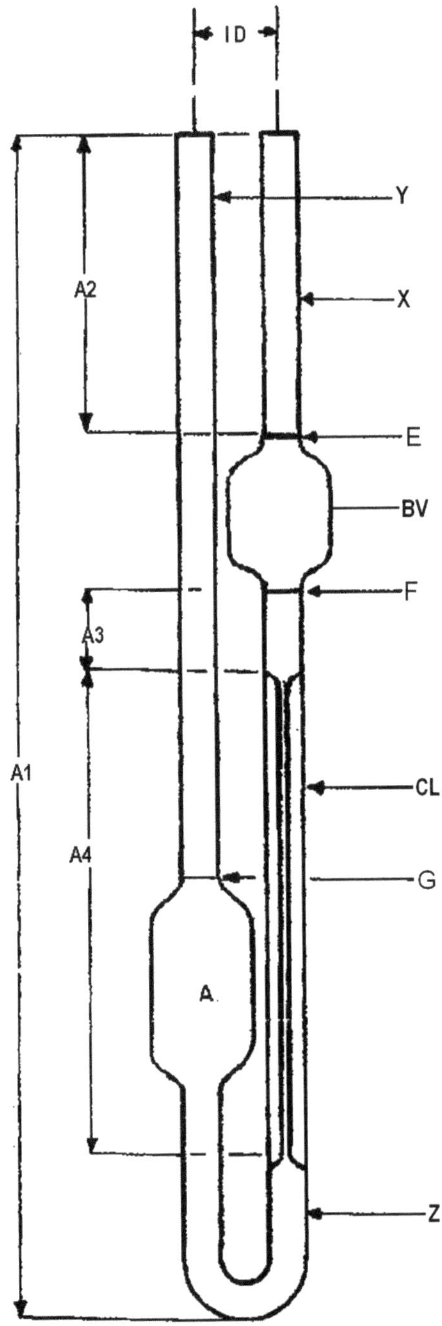

Fig. 3.5 BS/U-Tube viscometer

Table 3.5 Internal dimensions of ASTM viscometers

Size no.	Viscometer constant (mm^2/s^2)	Inside dia.of CL (mm) (±2 %)	Outside diameter of tubes in mm		Volume bulb BV (mL) (±5 %)	Vertical distance F to G (mm)	Outside Dia of bulbs A BV (mm)
			Y and Z	X			
A	0.003	0.50	8–9	6–7	5.0	91 ± 4	21–23
B	0.01	0.71	8–9	6–7	5.0	87 ± 4	21–23
C	0.03	0.88	8–9	6–7	5.0	83 ± 4	21–23
D	0.1	1.40	9–10	7–8	10.0	78 ± 4	25–27
E	0.3	2.00	9–10	7–8	10.0	73 ± 4	25–27
F	1.0	2.50	9–10	7–8	10.0	70 ± 4	25–27
G	3.0	4.00	10–11	9–10	20.0	60 ± 3	32–35
H	10.0	6.10	10–11	9–10	20.0	50 ± 3	32–35

3.3.7 Pinkevitch Viscometers

A typical Pinkevitch viscometer is shown in Fig. 3.7. The set consists of 11 viscometers covering the range from 0.6 to 17,000 mm^2/s.

3.3.7.1 Dimensions

The labelled dimensions of each viscometer, as shown in Fig. 3.7 are the same and can be found out from the relevant standard, for example, ASTM D446. Only internal dimensions vary, these are given in Table 3.7 [1].

3.3.7.2 Charging

To charge a Pinkevitch viscometer, invert the viscometer and apply suction/vacuum to the side arm O with finger on tube Y, with tube X immersed in the liquid sample to fill. Draw the sample to timing mark F. Mount the viscometer upright in the constant-temperature bath keeping tube L vertical. The tube Y is used to control the liquid flow.

3.3.8 Equilibrium Time

Allow the charged viscometer to remain in the bath long enough to reach the test temperature. Because this time will vary for different instruments, for different temperatures, and for different kinematic viscosities, establish a safe equilibrium time by trial (30 min should be sufficient except for very high kinematic

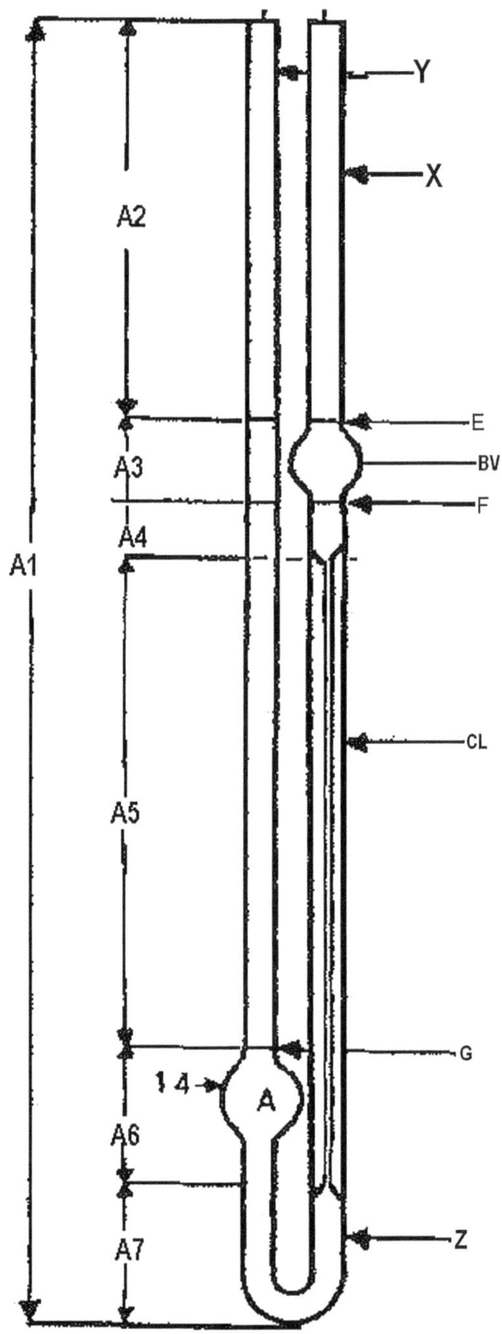

Fig. 3.6 BS/U/M miniature viscometer

Table 3.6 Internal dimensions of ASTM viscometers

Size no.	Viscometer constant (mm^2/s^2)	Inside dia. of CL (mm) (±2 %)	Outside dia. of tubes X, Y and Z (mm)	Volume of bulb BV (mL) (±5 %)
M1	0.001	0.20	6–7	0.50
M2	0.005	0.30	6–7	0.50
M3	0.015	0.40	6–7	0.50
M4	0.04	0.50	6–7	0.50
M5	0.1	0.65	6–7	0.50

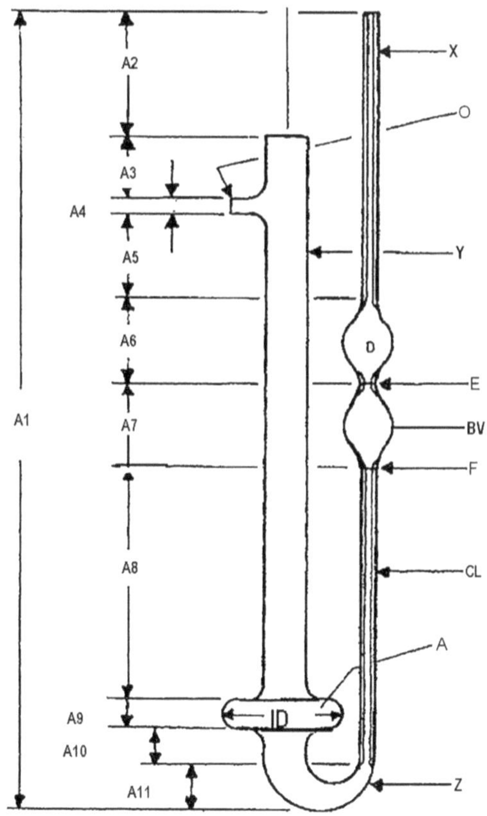

Fig. 3.7 Pinkevitch viscometer

viscosities). One bath is often used to accommodate several viscometers. Never add or withdraw a viscometer while any other viscometer is in use for measuring a flow time.

Table 3.7 Internal dimensions of ASTM viscometers

Size no.	Viscometer constant (mm^2/s^2)	Inside dia. of CL (mm) (±2 %)	Bulb volume (±5 %) (mL)	
			D	BV
0	0.0017	0.40	3.7	3.7
1	0.0085	0.60	3.7	3.7
2	0.027	0.80	3.7	3.7
3	0.065	1.00	3.7	3.7
4	0.14	1.20	3.7	3.7
5	0.35	1.50	3.7	3.7
6	1.0	2.00	3.7	3.7
7	2.6	2.50	3.7	3.7
8	5.3	3.00	3.7	3.7
9	9.9	3.50	3.7	3.7
10	17	4.00	3.7	3.7

3.3.9 Bringing the Sample up to the Timing Mark

Use vacuum to draw the sample through bulb BV to about 5 mm above upper timing mark E. In case of sample containing volatile constituents use pressure instead of vacuum. Release the vacuum or pressure as the case may be and allow the sample to flow under gravity.

3.4 Suspended Level Viscometers

3.4.1 Ubbelohde Viscometers

A typical Ubbelohde viscometer is shown in Fig. 3.8. The set consists of 16 viscometers covering the range from 0.3 to 100,000 mm^2/s.

3.4.1.1 Dimensions

The labelled dimensions of each viscometer, as shown in Fig. 3.8 are the same and can be found out from the relevant standard, for example, ASTM D446. Only internal dimensions vary, these are given in Table 3.8 [1].

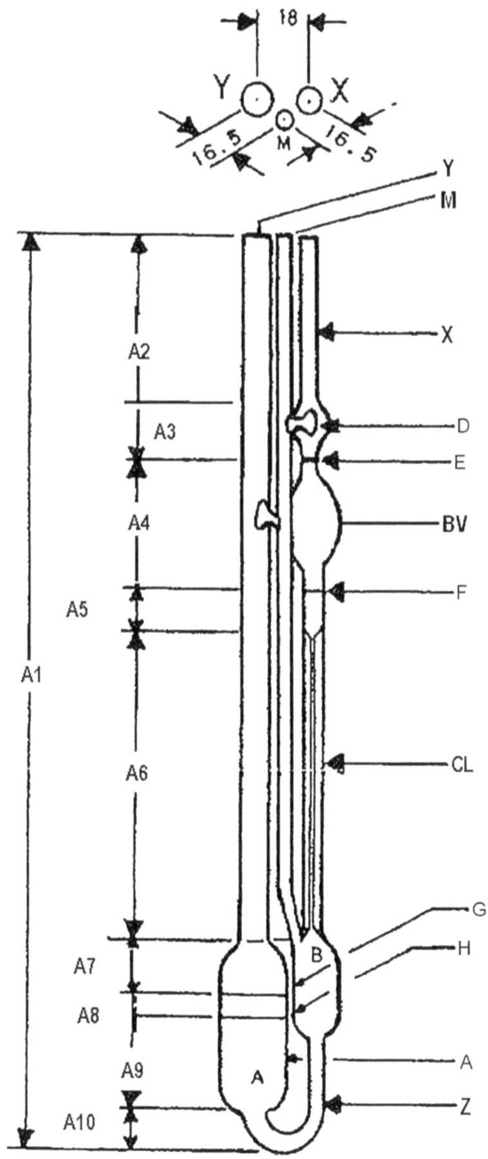

Fig. 3.8 Ubbelohde viscometer for transparent liquids

3.4.1.2 Charging

The Ubbelohde viscometer is charged by tilting the instrument about 30° from the vertical and pouring sufficient sample through the tube Y into bulb A so that when the viscometer is returned to the vertical, the meniscus is between fill marks G and H, and tube Z is completely filled without entrapping air. Mount the viscometer in the constant-temperature bath keeping the tube Y vertical.

Table 3.8 Internal dimensions of ASTM viscometers

Size no.	Viscometer constant (mm^2/s^2)	Inside dia. of Tube CL (mm) (±2 %)	Volume of bulb BV, (ml) (±5 %)	Inside dia. of tube Z (mm)
0	0.001	0.24	1.0	6.0
0C	0.003	0.36	2.0	6.0
0B	0.005	0.46	3.0	6.0
1	0.01	0.58	4.0	6.0
1C	0.03	0.78	4.0	6.0
1B	0.05	0.88	4.0	6.0
2	0.1	1.03	4.0	6.0
2C	0.3	1.36	4.0	6.0
2B	0.5	1.55	4.0	6.0
3	1.0	1.83	4.0	6.0
3C	3.0	2.43	4.0	6.0
3B	5.0	2.75	4.0	6.5
4	10	3.27	4.0	7.0
4C	30	4.32	4.0	8.0
4B	50	5.20	5.0	8.5
5	100	6.25	5.0	10.0

For more viscous liquids, the viscometer is inverted with tube Y placed in the sample. The vacuum/suction is applied to the tube X, closing the tube M by a finger or rubber stopper; draw sufficient sample into the tube Y such that after wiping Y clean and placing the viscometer in the constant-temperature bath, bulb A is filled between the marks G and H.

3.4.2 *Cannon Ubbelohde viscometer*

A typical Cannon-Ubbelohde viscometer is shown in Fig. 3.9. The set consists of 14 viscometers covering a range of 0.5 to 100,000 mm^2/s.

3.4.2.1 Dimensions

The labelled dimensions of each viscometer, as shown in Fig. 3.9 are the same and can be found out from the relevant standard, for example, ASTM D446. Only internal dimensions vary, these are given in Table 3.9 [1].

3.4.2.2 Charging

The Cannon-Ubbelohde viscometer is charged by tilting the instrument about 30° from the vertical and pouring sufficient sample through the tube Y into bulb A so that when the viscometer is returned to the vertical, the meniscus is between fill

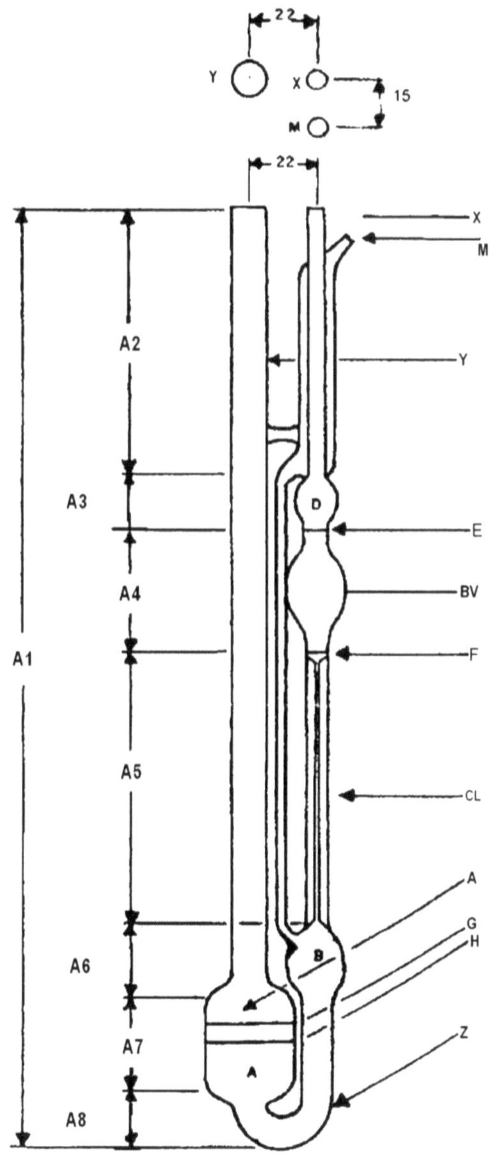

Fig. 3.9 Cannon-Ubbelohde viscometer

marks G and H, and tube Z is completely filled without entrapping air. Mount the viscometer in the constant-temperature bath keeping the tube Y vertical.

For more viscous liquids, the viscometer is inverted with tube Y placed in the sample. The vacuum/suction is applied to the tube X, closing the tube M by a finger or rubber stopper; draw sufficient sample into the tube Y such that after wiping Y clean and placing the viscometer in the constant-temperature bath, bulb A is filled between the marks G and H.

Table 3.9 Internal dimensions of ASTM viscometers

Size no.	Viscometer constant (mm^2/s^2)	Inside diameter of CL (mm) (±2 %)	Volume of bulb BV (mL) (±5 %)
25	0.002	0.31	1.5
50	0.004	0.44	3.0
75	0.008	0.54	3.0
100	0.015	0.63	3.0
150	0.035	0.78	3.0
200	0.1	1.01	3.0
300	0.25	1.26	3.0
350	0.5	1.48	3.0
400	1.2	1.88	3.0
450	2.5	2.25	3.0
500	8	3.00	3.0
600	20	3.75	3.0
650	45	4.60	3.0
700	100	5.60	3.0

Note: The cylindrical length of the bulb A is 30 mm for Cannon-Ubbelohde viscometer, while it is 80 mm for Cannon-Ubbelohde dilution viscometer.

3.4.3 Cannon-Ubbelohde Semi-Micro Viscometer

A typical Cannon-Ubbelohde semi-micro viscometer is shown in Fig. 3.10. The set consists of 12 viscometers covering the range from 0.4 to 20,000 mm^2/s.

3.4.3.1 Dimensions

The labelled dimensions of each viscometer, as shown in Fig. 3.10 are the same and can be found out from the relevant standard., for example, ASTM D446. Only internal dimensions vary, these, for example, are taken from ASTM D446 and given in Table 3.10 [1].

3.4.3.2 Charging

The Cannon-Ubbelohde Semi-Micro viscometer is designed both for semi-micro and dilution use, hence the filling marks G and H is omitted. The viscometer is charged by tilting the instrument about 30° from the vertical and pouring sufficient

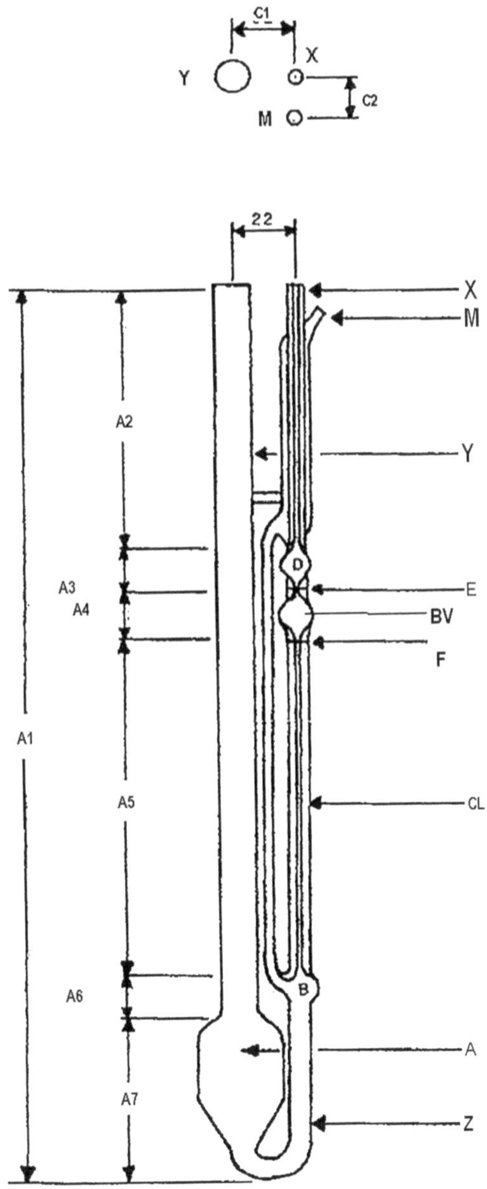

Fig. 3.10 Cannon-Ubbelohde semi-micro viscometer for transparent liquids

sample through the tube Y into bulb A so that when the viscometer is returned to the vertical, the bulb A as well as the tube CL is completely filled without entrapping air. Mount the viscometer in the constant-temperature bath keeping the tube Y vertical.

Table 3.10 Internal dimensions of ASTM viscometers

Size no.	Viscometer constant (mm^2/s^2)	Inside diameter of Tube CL (mm) (±2 %)	Volume of bulb BV (mL) (±5 %)	Inside dia. of tubes X, E, F and Z (mm)
25	0.002	0.22	0.30	1.2–1.4
50	0.004	0.25	0.30	1.2–1.4
75	0.008	0.30	0.30	1.2–1.4
100	0.015	0.36	0.30	1.2–1.4
150	0.035	0.47	0.30	1.2–1.4
200	0.1	0.61	0.30	1.4–1.7
300	0.25	0.76	0.30	1.5–1.8
350	0.5	0.90	0.30	1.8–2.2
400	1.2	1.13	0.30	2.1–2.5
450	2.5	1.40	0.30	2.4–2.8
500	8	1.85	0.30	2.7–3.1
600	20	2.35	0.30	3.7–4.0

3.4.4 BS/IP/SL (S) Viscometer

A typical BS/IP/SL (S) viscometer is shown in Fig. 3.11. The set consists of 9 viscometers covering the range from 1.05 to 10,000 mm^2/s.

3.4.4.1 Dimensions

The labelled dimensions of each viscometer, as shown in Fig. 3.11 are the same and can be found out from the relevant standard, for example, ASTM D446. Only internal dimensions vary, these are given in Table 3.11 [1].

3.4.4.2 Charging

Charge the BS/IP/SL(S) viscometer through tube Y with sufficient sample to fill bulb A, but not bulb B. The viscometer may be mounted vertically in the constant-temperature bath either prior to or following charging of the sample into the viscometer.

3.4.5 BS/IP/MSL Viscometer

A typical BS/IP/MSL viscometer is shown in Fig. 3.12. The set consists of 7 viscometers covering the range from 0.6 to 3,000 mm^2/s.

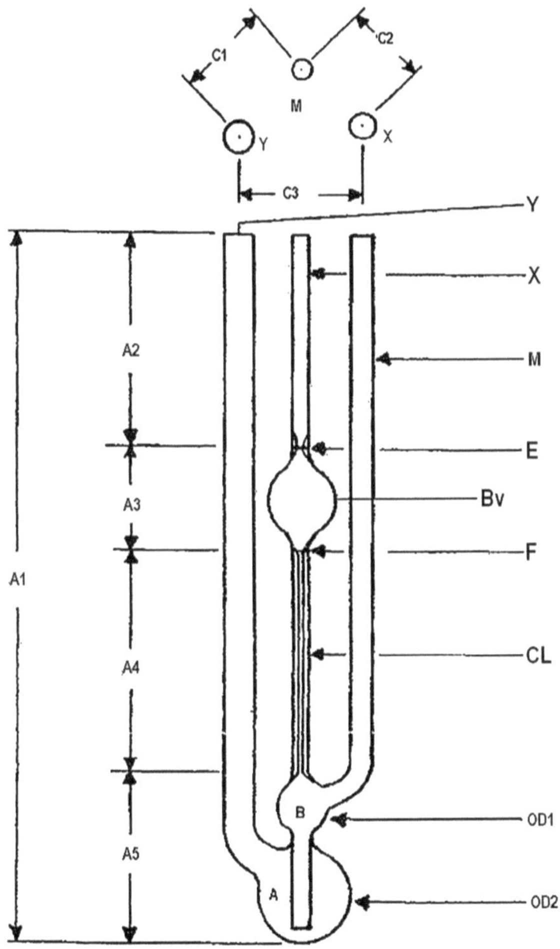

Fig. 3.11 BS/IP/SL (S) viscometer for transparent liquids

Table 3.11 Internal dimensions of ASTM viscometers

Size no.	Viscometer constant (mm^2/s^2)	Inside dia. of CL (mm) (±2 %)	Volume of bulb BV (ml) (±5 %)	Inside dia of tube X (mm)	Inside dia of Tube at E (mm)
1	0.0008	0.36	5.6	2.8–3.2	3
2	0.003	0.49	5.6	2.8–3.2	3
3	0.01	0.66	5.6	2.8–3.2	3
4	0.03	0.87	5.6	2.8–3.2	3
5	0.1	1.18	5.6	2.8–3.2	3
6	0.3	1.55	5.6	2.8–3.2	3
7	1.0	2.10	5.6	3.7–4.3	4
8	3.0	2.76	5.6	4.6–5.4	5
9	10.0	3.80	5.6	4.6–5.4	5

Fig. 3.12 BS/IP/MSL viscometer for transparent liquids

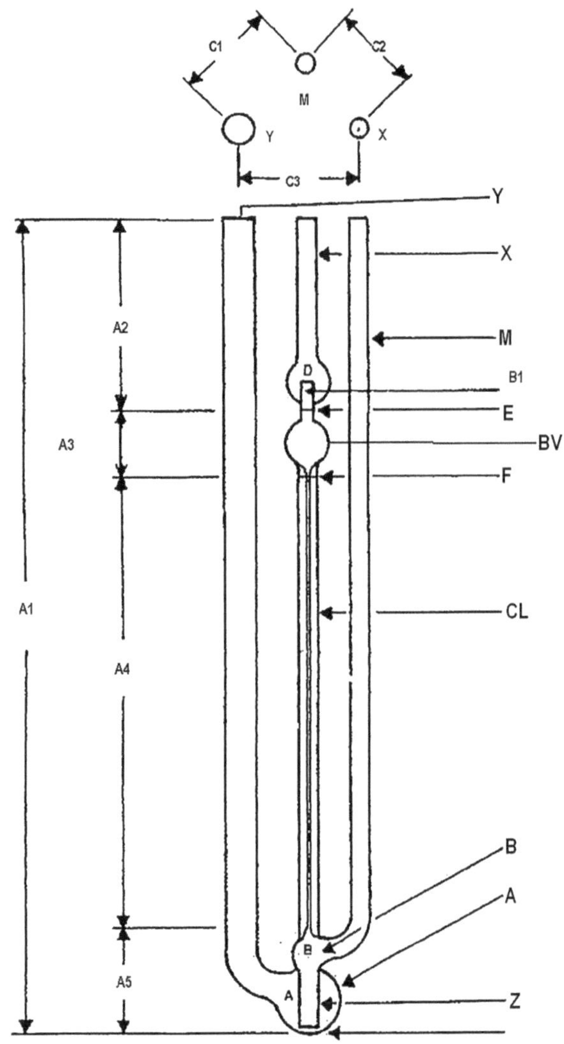

3.4.5.1 Dimensions

The labelled dimensions of each viscometer, as shown in Fig. 3.12 are the same and can be found out from the relevant standard, for example, ASTM D446. Only internal dimensions vary, these are given in Table 3.12 [1].

Table 3.12 Internal dimensions of ASTM viscometers

Size no.	Viscometer constant (mm^2/s^2)	Inside dia. of Tube CL (mm) (±2 %)	Volume of bulb BV (mL) (±5 %)	Inside dia. of tube X (mm)
1	0.003	0.35	1.2	4–6
2	0.01	0.45	1.2	4–6
3	0.03	0.62	1.2	4–6
4	0.1	0.81	1.2	4–6
5	0.3	1.10	1.2	4–6
6	1.0	1.45	1.2	4–6
7	3.0	1.98	1.2	4–6

3.4.5.2 Charging

Charge the BS/IP/MSL viscometer through tube Y with sufficient sample to fill bulb A, but not bulb B. The viscometer may be mounted vertically in the constant-temperature bath either prior to or following charging of the sample into the viscometer.

3.4.6 Fitz-Simons Viscometer

A typical Fitz-Simon viscometer is shown in Fig. 3.13. The set consists of 6 viscometers covering the range from 0.6 to 1,200 mm^2/s.

3.4.6.1 Dimensions

The labelled dimensions of each viscometer, as shown in Fig. 3.13 are the same and can be found out from the relevant standard, for example, ASTM D446. Only internal dimensions vary, these are given in Table 3.13 [1].

3.4.6.2 Charging

Charge the Fitz-Simons viscometer through tube Y with sufficient sample to fill bulb A, but not bulb B. The viscometer may be mounted vertically in the constant-temperature bath either prior to or following charging of the sample into the viscometer.

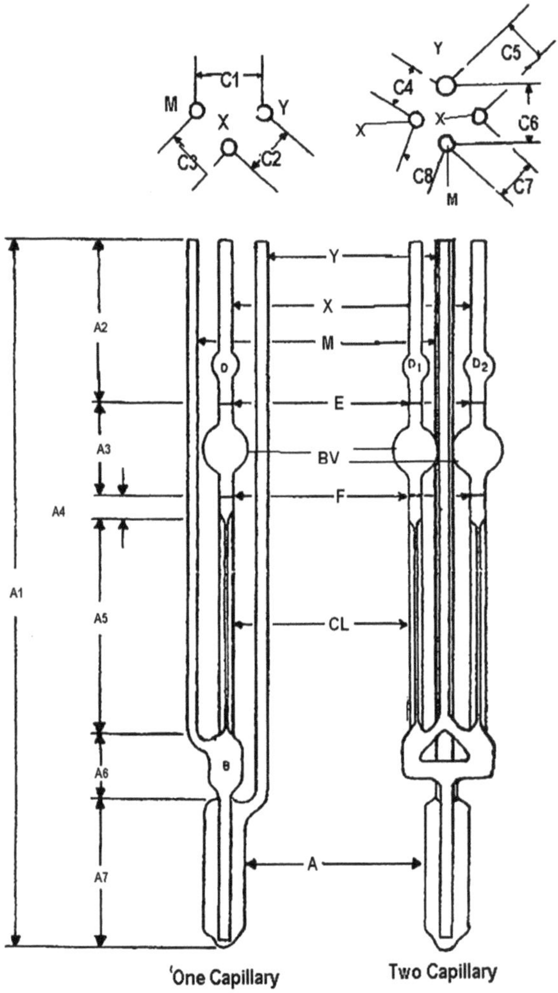

Fig. 3.13 Fitzsimons viscometers for transparent liquids

Table 3.13 Internal dimensions of ASTM viscometers

Size no.	Viscometer constant (mm^2/s^2)	Inside diameter of Tube CL (mm) (±2 %)	Volume of bulb BV (mL) (±5 %)
1	0.003	0.43	3.0
2	0.01	0.60	3.7
3	0.035	0.81	3.7
4	0.1	1.05	3.7
5	0.25	1.32	3.7
6	1.20	1.96	3.7

3.4.7 Atlantic Viscometer

A typical Atlantic viscometer is shown in Fig. 3.14. The set consists of 11 viscometers covering the range from 0.7 to 5,000 mm^2/s.

3.4.7.1 Dimensions

The labelled dimensions of each viscometer, as shown in Fig. 3.14 are the same and can be found out from the relevant standard, for example, ASTM D446. Only internal dimensions vary, these are given in Table 3.14 [1].

3.4.7.2 Charging

Permanently mount the Atlantic viscometer in the constant-temperature bath with the enlargement S resting on the top-split collar, and the lower end of capillary tube CL, 25 mm from the bottom of the bath. Pour the sample into a clean 50-mL beaker. Charge the viscometer by positioning the beaker and sample under tube Y so that it will be completely immersed in the sample. Slowly apply vacuum to tube X by turning the three-way stopcock O to vacuum. Draw the sample into the viscometer filling capillary CL, timing bulb BV, and partially filling upper bulb D. Close stopcock O, holding the sample in the viscometer. If only a small sample is available, a short length of rubber-tipped glass tubing can be placed in the beaker with the rubber against the bottom of capillary tube CL, and the sample drawn up as above.

3.4.8 Equilibrium Time

Allow the charged viscometer to remain in the bath long enough to reach the test temperature. Because this time will vary for different instruments, for different temperatures, and for different kinematic viscosities, establish a safe equilibrium time by trial (30 min should be sufficient except for very high kinematic viscosities). One bath is often used to accommodate several viscometers. Never add or withdraw a viscometer while any other viscometer is in use for measuring a flow time.

3.5 Reverse Flow Viscometers

3.5.1 Zeitfuchs Cross-arm viscometer

A typical Zeitfuchs Cross-arm viscometer is shown in Fig. 3.15. The set consists of 10 viscometers covering the range from 0.6 to 100,000 mm^2/s.

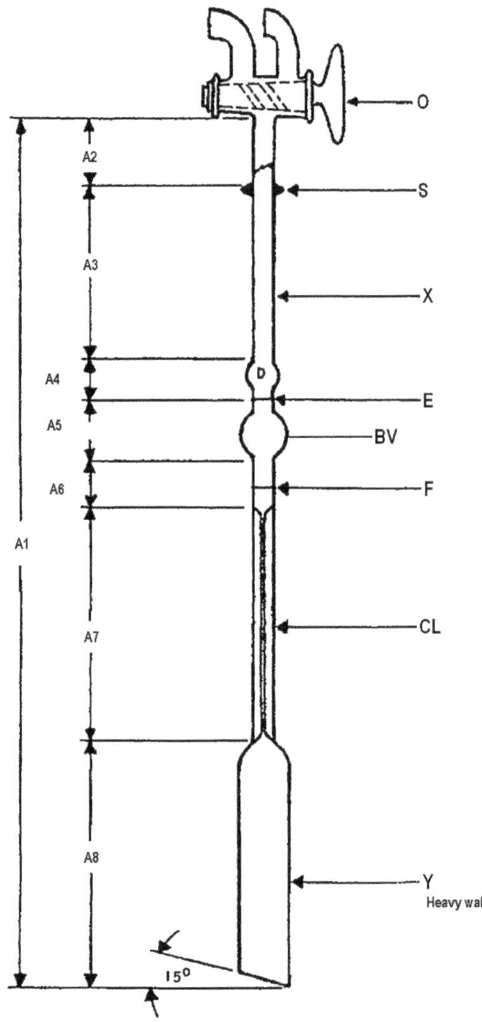

Fig. 3.14 Atlantic viscometer for transparent liquids

3.5.1.1 Dimensions

The labelled dimensions of each viscometer, as shown in Fig. 3.15 are the same and can be found out from the relevant standard., for example, ASTM D446. Only internal dimensions vary, these are given in Table 3.15 [1].

3.5.1.2 Charging

Mount the Zeitfuchs cross-arm viscometer in the constant-temperature bath, keep the tube X vertical. Introduce the test specimen through tube X, taking care not to

Table 3.14 Internal dimensions of ASTM viscometers

Size no.	Viscometer constant (mm^2/s^2)	Inside dia. of tube CL (mm) (±2 %)	Volume of bulb BV (mL) (±5 %)
0C	0.003	0.42	3.2
0B	0.005	0.46	3.2
1	0.01	0.56	3.2
1C	0.03	0.74	3.2
1B	0.05	0.83	3.2
2	0.1	1.00	3.2
2C	0.3	1.31	3.2
2B	0.5	1.48	3.2
3	1.0	1.77	3.2
3C	3.0	2.33	3.2
3B	5.0	2.64	3.2

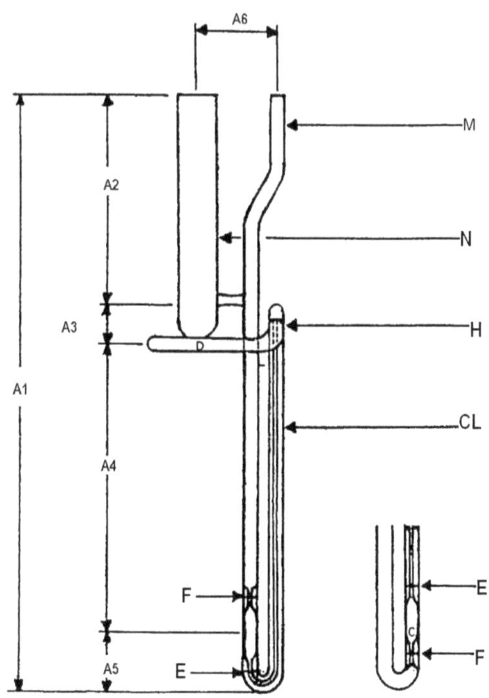

Fig. 3.15 Zeitfuchs Cross—arm viscometer for transparent and opaque liquids

wet the sides of tube X, into the cross-arm D until the leading edge stands within 0.5 mm of fill mark H on the siphon tube. The volume of the test specimen depends on the location of the fill mark H. When the flow time of the lower meniscus is being measured between timing marks E and F, the upper meniscus shall be in the horizontal cross-arm D, thus making the location of fill mark H critical.

Table 3.15 Internal dimensions of ASTM viscometers

Size no.	Viscometer constant (mm^2/s^2)	Inside dia. of tube CL (mm) (±2 %)	Length of CL (mm)	Lower bulb volume (ml) (±5 %)	Horizontal tube dia. (mm)
1	0.003	0.27	210	0.3	3.9
2	0.01	0.35	210	0.3	3.9
3	0.03	0.46	210	0.3	3.9
4	0.10	0.64	210	0.3	3.9
5	0.3	0.84	210	0.3	3.9
6	1.0	1.15	210	0.3	4.3
7	3.0	1.42	210	0.3	4.3
8	10.0	1.93	165	0.25	4.3
9	30.0	2.52	165	0.25	4.3
10	100.0	3.06	165	0.25	4.3

3.5.2 Cannon-Fenske Viscometer

A typical Cannon-Fenski viscometer is shown in Fig. 3.16. The set consists of 12 viscometers covering the range from 0.4 to 20,000 mm^2/s.

3.5.2.1 Dimensions

The labelled dimensions of each viscometer, as shown in Fig. 3.16 are the same and can be found out from the relevant standard., for example, ASTM D446. Only internal dimensions vary, these are given in Table 3.16 [1].

These viscometers work for transparent and opaque liquids.

3.5.2.2 Charging

Charge the Cannon–Fenske opaque viscometer by inverting it, immersing tube X in the liquid sample and applying suction to the tube Y. Draw liquid through tube X, filling bulb D to filling mark H. Wipe any excess sample off tube X and invert the viscometer to its normal position. Mount the viscometer in the constant temperature bath, keeping tube Y vertical. Close tube X with a rubber stopper or a short length of rubber tube with a screw clamp.

3.5.3 Lantz-Zeitfuchs Viscometer

A typical Lantz-Zeitfuchs viscometer is shown in Fig. 3.17. The set consists of 6 viscometers covering the range from 60 to 100,000 mm^2/s.

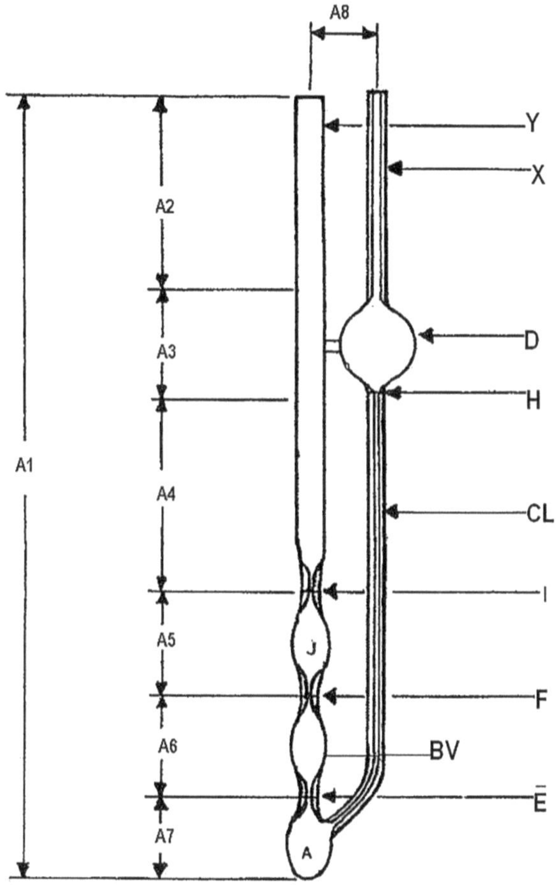

Fig. 3.16 Cannon-Fenske viscometer for transparent and opaque liquids

Table 3.16 Internal dimensions of ASTM viscometers

Size no.	Viscometer constant (mm^2/s^2)	Inside dia of tube CL (mm) (±2 %)	Inside dia. of tubes X, E, F & 1 (mm) (±5 %)	Volume of bulb BV (mL) (±5 %)	Volume of bulb D (mL)
25	0.002	0.31	3.0	1.8	11
50	0.004	0.42	3.0	2.1	11
75	0.008	0.54	3.0	2.1	11
100	0.015	0.63	3.2	2.1	11
150	0.035	0.78	3.2	2.1	11
200	0.1	1.02	3.2	2.1	11
300	0.25	1.26	3.4	2.1	11
350	0.5	1.48	3.4	2.1	11
400	1.2	1.88	3.4	2.1	11
450	2.5	2.20	3.7	2.1	11
500	8	3.10	4.0	2.1	11
600	20	4.00	4.7	2.1	13

3.5.3.1 Dimensions

The labelled dimensions of each viscometer, as shown in Fig. 3.17 are the same and can be found out from the relevant standard., for example, ASTM D446. Only internal dimensions vary, these, for example, are taken from ASTM D446 and given in Table 3.17.

These viscometers work for transparent and opaque liquids.

3.5.3.2 Charging

Mount the Lantz-Zeitfuchs viscometer in the constant-temperature bath, keeping tube X vertical. Introduce sufficient sample through tube X to completely fill bulb D, overflowing slightly into overflow tube K. If the sample is poured at a temperature above the test temperature, wait 15 min for the sample in the viscometer to attain bath temperature and add more sample to overflow slightly into tube K.

3.5.4 BS/IP/RF U-Tube Reverse Flow

A typical BS/IP/RF U-tube viscometer is shown in Fig. 3.18. The set consists of 11 viscometers covering the range from 0.6 to 300,000 mm^2/s.

3.5.4.1 Dimensions

The labelled dimensions of each viscometer, as shown in Fig. 3.18 are the same and can be found out from the relevant standard., for example, ASTM D446. Only internal dimensions vary, these, for example, are taken from ASTM D446 and given in Table 3.18.

3.5.4.2 Charging

Mount the BS/IP/RF viscometer in the constant temperature bath keeping the straight portion of the capillary tube CL vertical by using a plumb-line observed in two directions at right angles, or as stated in the certificate of calibration.

(1) Allow the viscometer to reach the bath temperature and then pour the filtered test sample into the filling tube X to a point just below the filling mark H avoiding wetting the glass above H.
(2) Allow the liquid to flow through the capillary tube CL, taking care that the liquid column remains unbroken, until it reaches a position about 5 mm below the filling mark I and arrest its flow at this point by closing the timing tube Y

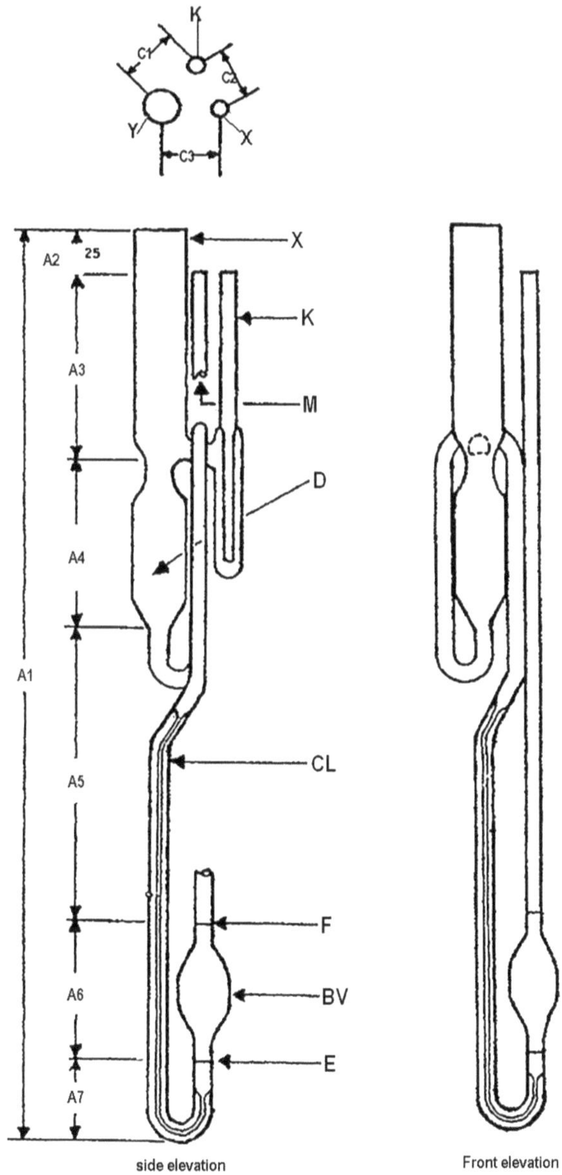

Fig. 3.17 Lantz-Zeitfuchs viscometer for transparent and opaque liquids

with a rubber bung. It is desirable that the rubber bung is fitted with a glass tube and stopcock so that one can apply a controllable, very slight excess pressure to tube Y.

(3) Add more liquid to the filling tube X to bring the oil surface to just below mark H. Allow the sample to reach the bath temperature and air bubbles to rise to the surface. The process requires at least 30 min.

Table 3.17 Internal dimensions of ASTM viscometers

Size no.	Viscometer constant (mm^2/s^2)	Inside dia. of tube CL (mm) (±2 %)	Length of tube CL (mm)	Volume of bulb BV (±5 %) (mL)
5	0.3	1.65	490	2.7
6	1.0	2.25	490	2.7
7	3.0	3.00	490	2.7
8	10.0	4.10	490	2.7
9	30.0	5.20	490	2.7
10	100.0	5.20	490	0.85

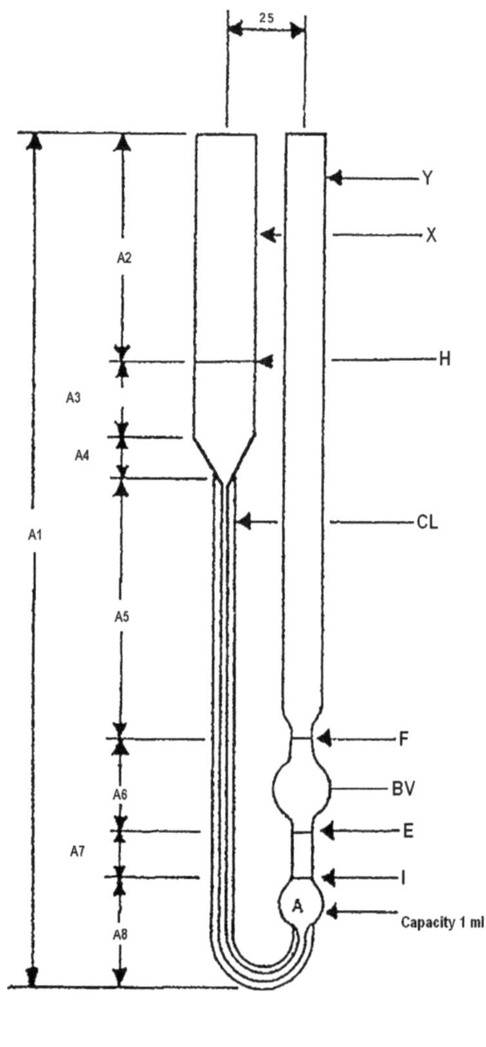

Fig. 3.18 BS/IP/RF U tube reverse flow viscometers opaque liquids

Table 3.18 Internal dimensions of ASTM viscometers

Size no.	Viscometer constant (mm^2/s^2)	Inside dia of tube CL (mm) (±2 %)	Length of tube CL (mm)	Inside dia. at E, F & I (mm)	Volume of bulb BV (mL) (±5 %)
1	0.003	0.51	185	3.0–3.3	4.0
2	0.01	0.71	185	3.0–3.3	4.0
3	0.03	0.93	185	3.0–3.3	4.0
4	0.1	1.26	185	3.0–3.3	4.0
5	0.3	1.64	185	3.0–3.3	4.0
6	1.0	2.24	185	3.0–3.3	4.0
7	3.0	2.93	185	3.3–3.6	4.0
8	10	4.00	185	4.4–4.8	4.0
9	30	5.5	185	6.0–6.7	4.0
10	100	7.7	210	7.7	4.0
11	300	10.0	210	10.0	4.0

(4) Gently manipulate the stopcock or bung closing the tube Y until the level of the liquid is arrested at mark I. The uppermost ring of contact of the sample with the glass should coincide with the bottom of mark I. Add sample to tube N until the uppermost ring of its contact with tube N coincides with the bottom of mark G.

3.5.5 Equilibrium Time

Allow the charged viscometer to remain in the bath long enough to reach the test temperature. Because this time will vary for different instruments, for different temperatures, and for different kinematic viscosities, establish a safe equilibrium time by trial (30 min should be sufficient except for the liquids of very high kinematic viscosity). One bath is often used to accommodate several viscometers. Never add or withdraw a viscometer while any other viscometer is in use for measuring a flow time.

3.5.6 Flow of Sample Through Capillary

For the Cannon-Fenske opaque and BS/IP/RF viscometers, remove the stopper in tubes X and Y respectively, and allow the sample to flow by gravity. For the Zeitfuchs cross-arm viscometer, apply slight vacuum to tube M (or pressure to tube X) to cause the meniscus to move over the siphon tube, and about 30 mm below the level of tube D in capillary CL; gravity flow is thus initiated. For the Lantz-Zeitfuchs viscometer, apply slight vacuum to tube M (or pressure tube X with tube K closed) until the lower meniscus is opposite the lower timing mark E; allow the sample to flow by gravity.

3.6 For all Samples

3.6.1 Sample for Charging Viscometers

If the sample is thought or known to contain fibbers or solid particles, filter through a 75–μm screen either prior to or during charging. To minimize the potential of particles passing through the filter from aggregating, it is recommended that the time lapse between filtering and charging be kept to a minimum.

References

1. ASTM D446-07 Standard specifications and operating instructions for glass capillary kinematic viscometers
2. ISO 3105-1994 Glass capillary kinematic viscometers—Specifications and operating instructions

Chapter 4
Rotational and Other Types of Viscometers

Abstract Capillary viscometers are essentially laboratory type instruments and do not find much place in measurement of viscosity in industry. Rotational viscometers such as Coaxial cylinders, concentric spheres, rotating disc, cone and plate, and coni-cylindrical, viscometers are commonly used in fields. Falling ball/piston viscometers together with rolling ball viscometers are also discussed. Cup and ball viscometer suitable for very small amount of liquid under test is given. Some viscometers, like Redwood No 1 and 2, Saybolt Universal and furol viscometers have been described. These are used in specific fields. Expressions of dynamic viscosity in terms of Saybolt universal viscometers are also given.

4.1 Introduction

Viscometers may be classified in four categories namely capillary viscometers, vibration viscometer, Rotational Viscometers and special purpose viscometers. Theory and details of Capillary viscometers is given in Chaps. 1 and 3; vibration viscometers have been discussed in Chap. 5. Rotational and special purpose viscometers are being taken in this chapter.

4.2 Rotational Viscometers

Rotational viscometers use the idea that the torque required to turn an object in a fluid is a function of the viscosity of that fluid. The measure of the torque required in rotating a disk or a body in a fluid at a known speed is proportional to the viscosity of the fluid.

S. V. Gupta, *Viscometry for Liquids*, Springer Series in Materials Science 194,
DOI: 10.1007/978-3-319-04858-1_4, © Springer International Publishing Switzerland 2014

Fig. 4.1 Couette viscometer

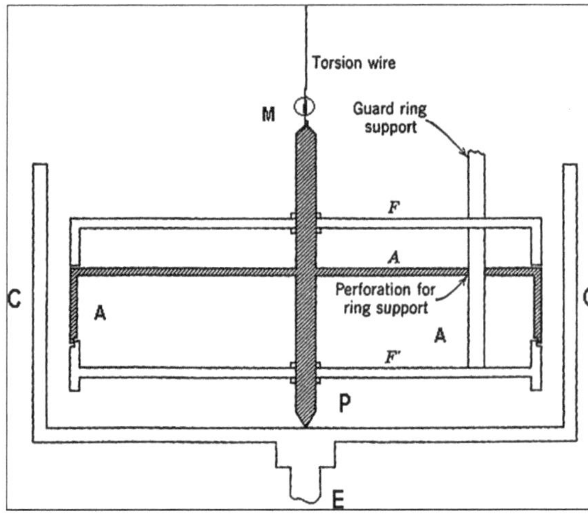

4.2.1 Coaxial Cylinders Viscometers

First practical rotational viscometer was devised by Couette in 1890 [1]. The line diagram of the Couette is shown in Fig. 4.1. It consisted of outer rotating cup C and a concentric cylinder A that is suspended by a torsion wire and resting on a point bearing P in the bottom of the cup C. Liquid whose viscosity was to be measured was filled in between the outer cup and inner cylinder. The inner cylinder A turned in proportional to the viscosity of the liquid till the torsion force of wire balanced the opposing viscous force.

The outer cylinder C supported by the spindle E rotates at a fixed speed. The inner cylinder A is suspended from a torsion wire B and centered by the pivot P. Both edges of the suspended cylinder are protected by two guard rings F and F′ concentric with the cylinder are rigidly held in position. The rings are perforated with end plates to prevent the liquid, between the two cylinders, from rotating. The deflection of the inner cylinder is measured by a mirror M attached to the torsion wire. A parallel beam of light reflecting from the mirror, measures the deflection of the torsion wire.

Suppose the outer cup of radius b is rotated with a constant angular velocity Ω. If there is no slip then the liquid layer in contact of outer cup will have an angular velocity Ω that in contact with inner cylinder will be stationary. That the velocity of the layers with increasing radius r from the central axis will be increasing from 0 at $r = a$, to Ω at $r = b$. If angular velocity of a layer at radius r is ω, the gradient in velocity is $\frac{d\omega}{dr}$. The viscous force exerted on the cylindrical layer of radius r and length h will be equal to the product of area of cylindrical layer, viscosity of the liquid and linear velocity gradient $r\frac{d\omega}{dr}$, the product is equal to

$$2\pi rh \cdot \eta \cdot r\frac{d\omega}{dr} \tag{4.1}$$

Giving the moment of the couple as

$$2\pi r^3 h \cdot \eta \cdot \frac{d\omega}{dr} \tag{4.2}$$

If there is no acceleration then couple of the torsion wire ΔG will be given by

$$2\pi h \cdot \eta \cdot r^3 \frac{d\omega}{dr} = \Delta G \tag{4.3}$$

Integrating with respect of r in between b and a and rearranging we get

$$G = 4\pi h \cdot \eta \cdot \Omega . a^2 \frac{b^2}{(b^2 - a^2)} = K\eta\Omega \tag{4.4}$$

Here k is a constant for any given apparatus.

By measuring G for given value of Ω, one can measure the viscosity of the liquid intervening between the two cylinders.

In deriving the formula the length of the cylinders was assumed to be infinite. In practice both cylinders are of finite length. Hence some account of the end effects is to be taken. The elimination of end effects is not necessary when relative viscosity is to be measured with that of the standard liquid. For absolute measurements guard rings are provided to almost eliminate the end effect. This apparatus is more often used for low viscosity liquids. The torque can be conveniently used by measuring the deflection of the inner cylinder suspended by a thin torsion wire.

Equation (4.4) is valid only up to the certain angular velocity of the moving cylinder [1] i.e. till the liquid flow is laminar in between the two cylinders. For this purpose a dimension less number known as Reynold number R_e is defined as follows:

$$R_e = \frac{\rho v L_1}{\eta} \tag{4.5}$$

Here

ρ is density of liquid,

v is average velocity,

L_1 is the distance over the velocity is averaged and

η is the dynamic viscosity of the liquid.

For the arrangement under discussion, in case outer cylinder is moving with angular velocity Ω, then linear velocity $v = b\Omega$, $(b-a)$ is the distance over which velocity is averaged and v is ratio of η/ρ is the kinematic viscosity, giving R_e as

$$R_e = \frac{b\Omega(b - a)}{v} \tag{4.6}$$

Couette observed that the flow is laminar till the Reynold number R_e as given above reaches about 1900. In some cases this value can be stretched to 32 000.

The Equation (4.4) is equally applicable if the inner cylinder rotates with constant angular velocity Ω and outer cup is fixed and suspended from the torsion wire. Such an arrangement is more suitable for more viscous liquids.

The maximum angular velocity Ω_c with which inner cylinder may rotate till the laminar flow is maintained as derived by Merrington [2] is given by a relation

$$\frac{\Omega_c^2 a^2 d^3}{0.5(a+b)v} = \frac{\pi^4 f}{0.0571 f^2 + 0.00056} \tag{4.7}$$

Here $d = b{-}a$, $f = 1{-}0.652d/a$ and d/a is small.

4.2.2 Concentric Spheres Viscometer

When a sphere of radius a rotates in a liquid with angular velocity Ω in a liquid of viscosity η contained within a fixed concentric sphere of radius b, then it will experience a couple G due to viscous forces given as:

$$G = -8\pi\eta\Omega\frac{a^3b^3}{b^3 - a^3} \tag{4.8}$$

For an accuracy of 0.1 % it is sufficient if the radius of the outer fixed sphere b is 10 times the radius a of the inner rotating sphere. In that case the G becomes

$$G = -8\pi\eta\omega a^3 \tag{4.9}$$

A modification of this principle was used by Ungar [3–5], in which a suspended hemisphere dips into the liquid contained in a fixed hemi-spherical cup. If the liquid level coincides with the centre of curvature of both the spheres the retarding couple is only half that given by (4.9).

4.2.3 Rotating Disc Viscometer

A disc of known dimensions is rotated with a constant angular velocity within a given liquid, needs a torque G to maintain a constant angular velocity Ω and is proportional to the viscosity of the liquid. So the system works as a viscometer. However it is rather difficult to get an accurate expression between viscosity η, Ω and other dimensions of the disc. This method is, therefore, not used for absolute determinations of viscosity. For a thin disc of radius a rotating in an infinite Newtonian liquid, Schlichting [6] has given the expression for the torque G to maintain the angular velocity Ω as

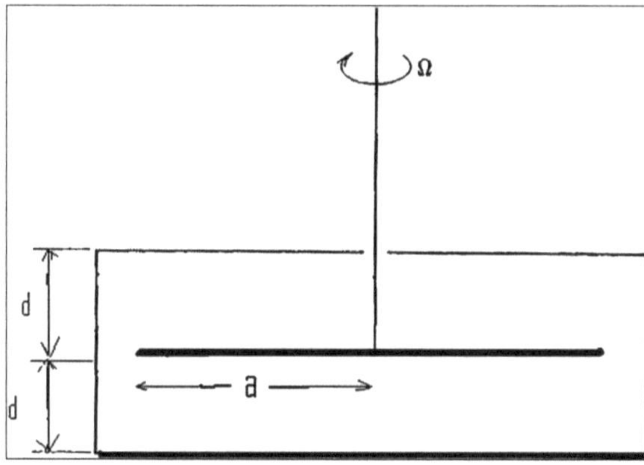

Fig. 4.2 Disc viscometer

$$2G = 0.616\pi\rho a^4 (\Omega^3 v)^{3/2} \tag{4.10}$$

The radius, *a*, of the disc is large enough to neglect the edge effects.

Here v is the kinematic viscosity and ρ is the density of the liquid.

In the case of an infinitely thin disc rotating between the two equally placed fixed plates the viscosity η is given as

$$\eta = \frac{Gd}{\pi a^4 \Omega} \tag{4.11}$$

The distance of each fixed plate from the disc is *d*.

To account for the end correction, Maxwell [7] derived a correction and the expression for η is given by

$$\eta = \frac{Gd}{\pi a^4 \Omega} + \frac{G}{8a^3 \Omega \ln(2)} \tag{4.12}$$

The necessary condition is that $d/a \ll 1$.

The Fig. 4.2 shows the schematic diagram of a disc viscometer.

4.2.4 Cone and Plate Viscometer

For a cone of large vertex angle rotating on a plate with a constant angular velocity Ω, the viscosity η of the liquid in between the plate and cone can be determined by measuring the torque G and other dimensions of the cone. The expression for viscosity in terms of dimensions of the cone and torque G is given as

$$\eta = \frac{3\alpha G}{2\pi a^3 \Omega} \tag{4.13}$$

Fig. 4.3 Conical viscometer

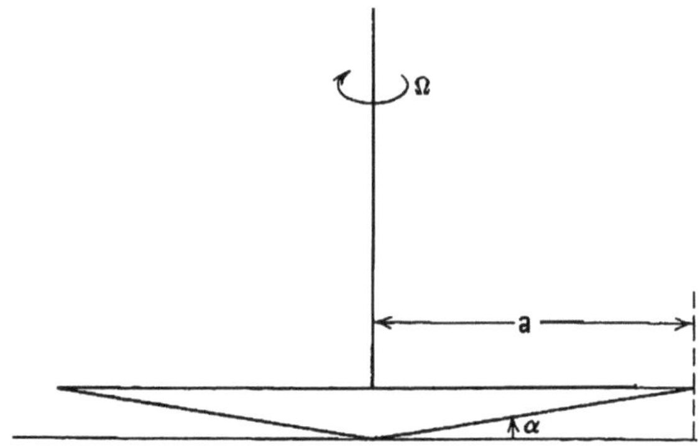

where α is the angle which the conical surface makes with the plate and a is the radius of the base of the cone. For larger values of α, the expression for viscosity η for a Newtonian liquid is given as

$$\eta = G\frac{3\{(\sin\alpha/\cos^2\alpha) - \ln\,\tan(\pi/4 - \alpha/2)\}}{4\pi a^3\Omega} \tag{4.14}$$

The diagram of such a viscometer is shown in Fig. 4.3.

4.2.5 Coni-Cylindrical Viscometer

A conic-cylindrical viscometer is schematically shown in Fig. 4.4. The advantage of such a viscometer is that mean rate of shear in the cylindrical annulus and in the conical portion is about the same and the end effect is nearly eliminated. The relation between viscosity η, measured torque G on rotating system and other dimensions of viscometer is

$$\frac{G}{\Omega} = \frac{4\pi\eta h}{(1/R_b^2 - 1/R_c^2)}\left\{1 + \frac{\Delta h}{h}\right\} \tag{4.15}$$

where Δh is given as

$$\Delta h = \frac{h_c}{6\cos^2\alpha_1}\left\{\frac{1}{\tan^2\alpha_1} - \frac{1}{\tan^2\alpha_2}\right\}\left\{\int_{\alpha_1}^{\alpha_2}\frac{d\alpha}{\sin^2\alpha}\right\}^{-1} \tag{4.16}$$

Here h and h_c are respectively the heights of cylinder and cone of the viscometer.

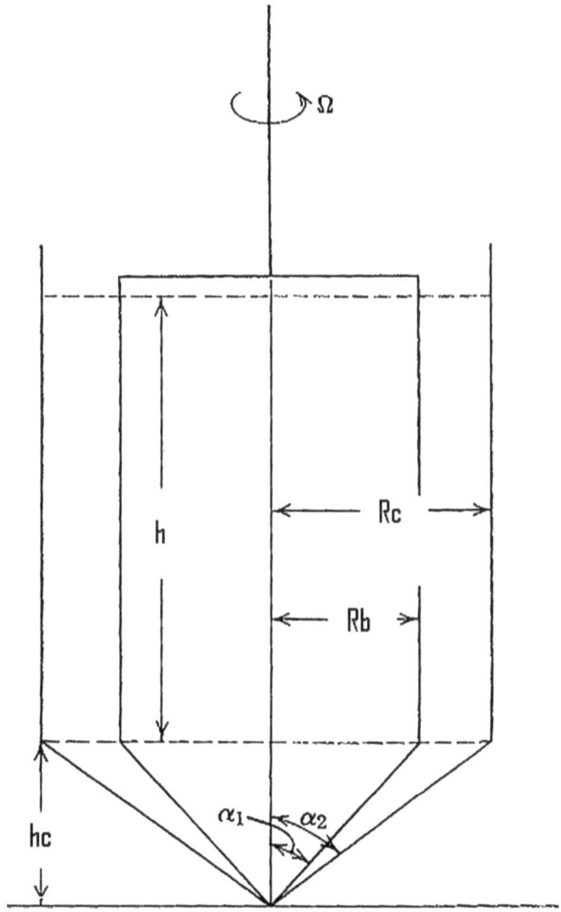

Fig. 4.4 Coni-cylindrical viscometer

4.3 Falling Ball/Piston Viscometers

Stokes' law is the basis of the falling sphere viscometer, in which the fluid is stationary in a vertical glass tube. A sphere of known size and density is allowed to descend through the liquid. If correctly selected, it reaches terminal velocity. The terminal velocity is the maximum velocity which a body will attain during a free fall in a medium. The terminal velocity is obtaining by measuring time it takes to pass two marks on the tube. Electronic sensing can be used for opaque fluids. Knowing the terminal velocity, the size and density of the sphere, and the density of the liquid, Stokes' law can be used to calculate the viscosity of the fluid. A series of steel ball bearings of different diameters are normally used in the classic experiment to improve the accuracy of the calculation. The school experiment uses glycerin as the fluid, and the technique is used industrially to

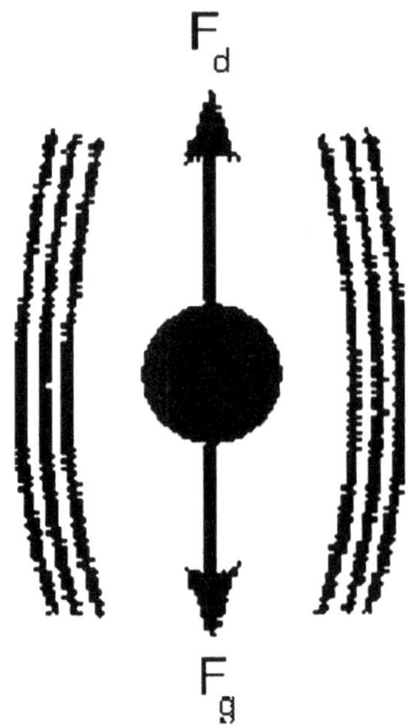

Fig. 4.5 Falling sphere through a liquid

check the viscosity of fluids used in various processes. It includes many different oils, and polymer liquids such as solutions (Fig. 4.5).

In 1851, George Gabriel Stokes derived an expression for the frictional force (called drag force) exerted on spherical objects with very small Reynolds numbers (e.g., very small particles) in a continuous viscous fluid. The equation is

$$F = 6\pi r\eta V \tag{4.17}$$

where:

F is the frictional force,
r is the radius of the spherical object,
η is the fluid viscosity, and
V is the velocity of the sphere.

If the particles are falling in the viscous fluid by their own weight, then a terminal velocity, also known as the settling velocity, is reached when this frictional force (4.17) combined with the buoyant force ($4\pi r^3 g\rho_f/3$) exactly balances the gravitational force $4\pi r^3 g\rho_s/3$. The resulting terminal velocity is given by:

$$V_t = \frac{2}{9} \cdot \frac{r^2 g(\rho_s - \rho_f)}{\eta} \tag{4.18}$$

where:

V_t is the terminal velocity (m/s) of sphere (vertically downwards if $\rho_s > \rho_f$ and upwards if $\rho_s < \rho_f$)

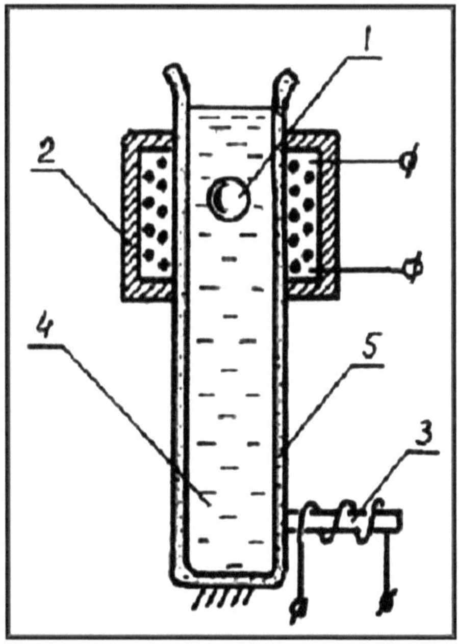

Fig. 4.6 Falling ball viscometer

r is radius of the sphere (m),
g is gravitational acceleration (m/s^2),
ρ_s is density of the sphere (kg/m^3),
ρ_f is density of the fluid (kg/m^3), and
η is (dynamic) fluid viscosity (Pa·s).

Note that Stokes law is valid for small Reynolds number. A limiting factor on the validity of this result is the roughness of the sphere being used (Fig. 4.6).

The size of the sphere depends on the viscosity. The minimum viscosities which can be measured by steel ball are given as

Diameter of the ball	2/32	3/32	4/32	6/32	8/32	12/32	16/32 inch
Viscosity	1000	2500	4000	10000	17000	38000	68000 mPa·s

4.3.1 *Falling Ball Viscometer*

A falling-ball viscometer is comprised of a tube with the liquid under-test, piston or ball inside the tube, an electrical magnet, and a magnetic switch. The piston or ball is made of a ferromagnetic material. It is first lifted to the top of the tube by the magnet and is then permitted to drop to the bottom of the tube under the force of gravity. The moment when the piston or ball reaches the bottom is sensed by a magnetic switch. The time required to pass the length of the tube is proportional to the viscosity. The average time of three tests is taken; the result is converted into a viscosity value using a simple formula given above in (4.18).

However it has been found that the measured value of the viscosity is slightly more than the true value. The one of the relations between measured η_{meas} and true η_{true} viscosities is:

$$\eta_{true} = \eta_{meas}\left\{1 - 2.104\frac{r}{R} + 2.09\frac{r^3}{R^3} - 0.95\frac{r^5}{R^5}\right\} \tag{4.19}$$

Another relation is

$$\eta_{true} = \frac{\eta_{meas}}{\left\{1 + 2.4\frac{r}{R}\right\}\left\{1 + 3.3\frac{r}{h}\right\}} \tag{4.20}$$

Here R and h are the radius and height of the cylinder containing the liquid through which the sphere of radius r is falling freely.

4.3.2 Falling Piston Viscometer

Falling piston viscometer is also known as the Norcross viscometer after its inventor, Austin Norcross. The principle of viscosity measurement in this rugged and sensitive industrial device is based on a piston and cylinder assembly. The piston is periodically raised by an air lifting mechanism, drawing the material being measured down through the clearance (gap) between the piston and the wall of the cylinder into the space which is formed below the piston as it is raised. The assembly is then typically held up for a few seconds, then allowed to fall by gravity, expelling the sample out through the same path that it entered, creating a shearing effect on the measured liquid, which makes this viscometer particularly sensitive and good for measuring certain liquids. The time of fall is a measure of viscosity, with the clearance between the piston and inside of the cylinder forming the measuring orifice. The viscosity controller measures the time of fall and displays the resulting viscosity value. The controller of Norcross viscometer can display the time-of-fall value to cup in seconds (known as efflux cup), Saybolt universal second (SUS) or in centipoise.

Industrial use is popular due to simplicity, repeatability, low maintenance and longevity. This type of measurement is not affected by flow rate or external vibrations. The principle of operation can be adapted for many different conditions, making it ideal for process control environments.

4.4 Rolling Ball Viscometer

A modification of the straight falling sphere viscometer is a rolling ball viscometer in which the time, for a ball rolling down a slope whilst immersed in the test fluid, is measured.

The speed with which a sphere rolls down a cylindrical tube filled with a fluid and the tube is inclined at a fixed angle to the horizontal gives a measure of the viscosity of the fluid. However the mathematical derivation is quite complicated, especially when the diameter of the ball is close to that of the tube. Hubbard and Brown [8] carried out the theoretical and practical investigations of the rolling ball system and deduced a general equation for the motion of a Newtonian fluid for the stream line flow. The equation can be reduced for any one instrument under constant conditions as

$$\eta = k(\rho_l - \rho_s)/V \tag{4.21}$$

Here k is a constant of the instrument depending upon the diameters of the tube and ball, inclination of the tube viscosity of the liquid and temperature etc. V is the translational velocity of the ball; ρ_s is the density of the ball and ρ_l is density of the liquid. This can be further improved by using a patented V plate which increases the number of rotations for a given distance traveled, allowing smaller more portable devices. This type of device is also suitable for ship board use. The account of the rolling ball viscometers has been given by Hersay [9] and Flowers [10].

4.4.1 Measurement with Rolling Ball Viscometer

The rolling and sliding movements of a ball through the sample liquid are timed in an inclined cylindrical measuring tube. The sample viscosity is correlated to the time needed by a ball to traverse a definite distance. By turning the measuring tube upside down again the return of the ball may also be used for an additional measurement. The test results are given in millipascal seconds (mPa·s) a sub multiple of SI unit for dynamic viscosity.

4.5 Torsion Viscometer

A device for measuring viscosity of thermoplastic polymer resins consists of a tall cylindrical cup of inner radius R_2 which holds the resin, and a cylindrical rod of radius R_1 inserted downward through most of the length of the container, which is far enough from the bottom that shear below the rod is negligible. The rod is attached to a motor which turns it at a constant rate, and the cup is attached to a torque meter. The device measures the torques for several rotational speeds of the rod.

The torsion viscometer Gallenkamp type makes viscosity measurements following the principle that a rotating cylinder immersed in a fluid is subject to a viscous drag. The result is shown on disk with graduated scale.

Forgacs described an improved version of the torsion viscometers for polymers [11] to permit determination of dynamic viscosity and dynamic modulus of solid and liquid polymers. Sample temperature is automatically controlled within less than 1 °C to any temperature in the range of -80 to $+100$ °C. Forced oscillation of the pendulum at frequencies from 3×10^{-6} to 3 Hz or higher is possible, also, freely decaying oscillation and creep testing. The viscometer is extremely sensitive and can detect angular deflection of $1''$ of arc. A stable photoelectric angle detector is combined with a thermoelectric temperature control system, which dispenses with vibration inducing moving fluids. This makes possible measurements with shear strain amplitudes as low as 10^{-6} to 10^{-7}, and results in improved low frequency angular measurement accuracy, limited by the ambient level of room vibrations.

4.6 Oscillating Piston Viscometer

Sometimes referred to as electromagnetic viscometer or EMV viscometer, was invented at Cambridge Viscosity (Formally Cambridge Applied Systems) in 1986 [12]. The sensor, Fig. 4.7, comprises a measurement chamber and magnetically influenced piston. Measurements are taken whereby a sample is first introduced into the thermally controlled measurement chamber M where the piston P resides. Electronics drive the piston into an oscillatory motion within the measurement chamber with a controlled electro-magnetic EM. A shear stress is imposed on the liquid under-test (or gas) due to the piston travel D and the viscosity is determined by measuring the travel time of the piston. The construction parameters for the annular spacing between the piston and measurement chamber, the strength of the electromagnetic field, and the travel distance of the piston are used to calculate the viscosity according to Newton's Law of Viscosity.

The oscillating piston viscometer technology has been adopted for small sample viscosity and micro-sample viscosity testing in laboratory applications. It has also been adapted to measure high pressure viscosity and high temperature viscosity measurements in both laboratory and process environments. The viscosity sensors have been scaled for a wide range of industrial applications such as small size viscometers for use in compressors and engines, flow-through viscometers for dip coating processes, in-line viscometers for use in refineries, and hundreds of other applications. Improvements in sensitivity from modern electronics, is stimulating a growth in oscillating piston viscometer popularity with academic laboratories exploring gas viscosity.

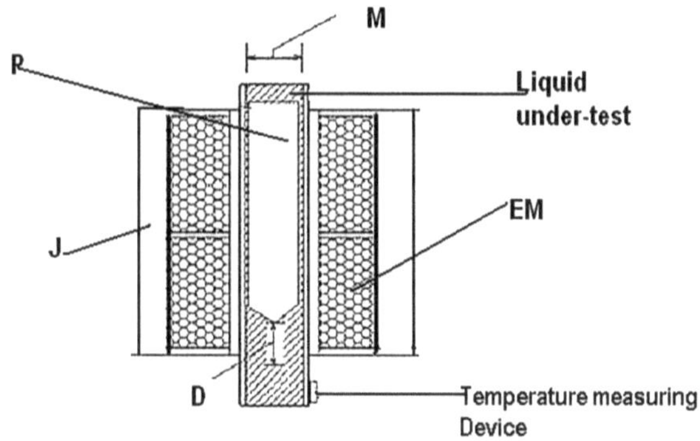

Fig. 4.7 Oscillating piston viscometer

4.7 Michell Cup and Ball Viscometer

If two planes and parallel surfaces separated by a layer of liquid are brought together by application of forces normal to the surfaces, liquid will tend to flow from the centre to the outside edges. The flow will be almost laminar if the separation of the planes is small in relation to the area and the movement is small. Reynolds calculated that the force F applied in this way to two circular discs of radius a, will reduce the initial separation H_o to H_1 in time t is related to viscosity η as

$$t = \frac{3\pi\eta a^4}{4F}\left\{\frac{1}{H_1^2} - \frac{1}{H_o^2}\right\} \tag{4.22}$$

This principle has been used in the cup-and-ball viscometer. The viscometer is outlined in Fig. 4.8.

4.7.1 Construction

A cup is attached to a hollow handle. In the hollow of the handle one can insert a thermometer for temperature measurement. A steel spherical ball of one inch (25.4 mm) diameter provides another surface. The curvature of the cup is kept the same as that of the spherical ball. Three small studs are attached to cup. The studs are accurately grounded to project about 0.01 mm, so that a space of uniform thickness of 0.01 mm is left in-between the surfaces of the ball and the cup. The liquid under test is filled in between the space. Only a few μl of liquid is required for the purpose.

Fig. 4.8 Cup-and-ball viscometer

4.7.2 Working

There are two options for its working. One is that liquid is first placed on the ball of sufficient thickness so that when cup is place over it, a small amount of liquid oozes out. That is the initial thickness of the liquid film should be larger than 0.01 mm so that there is no air gap when the cup is placed over it. The cup is lifted with the handle and time the ball takes to separate is noted. Normal force F to increase the separation is provided by the weight of the sphere. Alternatively, ball is placed in a vessel full of liquid under test and the cup is placed over it smoothing without any air bubble. The handle is again lifted and the time which the ball takes to leave the cup is noted. The time taken in each case will be proportional to the viscosity of the liquid. Viscosity is then the time taken to separate ball from the handle divided by instrument constant k.

The instrument is calibrated with a liquid of known viscosity η and instrument constant k is calculated by

$$k = \frac{t}{\eta} \tag{4.23}$$

An accuracy of 4 % is claimed from such instruments.

4.8 VROC

Rheo Sense introduced the micro-fabricated Rectangular Channel Viscometer (VROC). The VROC stands for Viscometer/Rheometer-on-a-Chip and it measures viscosity from the pressure drop while a test liquid flows through a rectangular slit, a well known scientific application [13].

4.8.1 Physical Structure

The chip consists of a rectangular slit that is formed with glass and an array of Silicon pressure sensors. The chip has a unique rectangular slit that mitigates perturbations at the entrance, sides and exit. This directly contributes to the enhanced accuracy of the sensor and ability to measure with precision in the presence of even high shear.

When the test sample is pumped to flow through the slit channel, the monolithic pressure sensor array measure pressure at different positions. Pressure linearly decreases and its slope *m* is calculated from pressure versus position graph. In this case the flow disturbance is negligible.

4.8.2 Results Analysis

The measured pressure as a function of position would be linear if a fully developed flow is ensured in the rectangular slit channel. From the slope, the wall shear stress (τ) is given as:

$$\tau = -m\frac{wh}{2(w+h)} \tag{4.24}$$

and apparent shear γ_{app} as

$$\gamma_{app} = \frac{6Q}{wh^2} \tag{4.25}$$

Giving viscosity η as

$$\eta = \frac{\tau}{\gamma_{app}} = \frac{mw^2h^3}{12(w+h)Q} \tag{4.26}$$

Here

w is width of the channel (slit) and h is channel (slit) depth.

For the Newtonian liquids, the above formula is sufficient. However, for non-Newtonian liquids the apparent shear (γ_{app}) does not remain equal the true shear rate (γ_{true}). For the rectangular slit flow, the true shear rate is calculated by applying Weissenberg-Rabinowitsch correction. To ensure accuracy, software application is supplied along with the sensor which applies this correction for non-Newtonian measurements:

The correction is:

$$\gamma_{true} = \frac{\gamma_{app}}{3}\left\{2 + \frac{d\log(\gamma_{app})}{d(\log(\tau)}\right\} \tag{4.27}$$

4.8.3 *Advantage of Small Gap*

The small gap of the flow channel provides significant advantages, such as:

1. Accurately measures the viscosity of small samples at extremely high shear rates, a service that normal rheometers cannot provide.
2. With the small gap, mimic the lubrication, high speed coating, and inkjet process.

4.9 Viscometers for Specific Field

Some viscometers have been developed keeping in view the need of a specific industry. In the oil industry the modified form of essentially capillary viscometers with very short length of capillary is quite common. These are Redwood No. 1 and No. 2 principally used in Great Britain, Saybolt Universal and Saybolt Furol, which are used in America. In Germany similar viscometer is named as Engler viscometer.

All the five aforesaid instruments are used more or less in the same way, but the finer constructional details are different. Essentially all have a cup with a small capillary (orifice) placed inside a thermo-control chamber, filled with oil under test and time of flow of a specified volume of oil is measured. Because of the comparatively smaller capillary length, flow of oil does not strictly obey Poiseuille's law; hence flow time and viscosity are not related by a simple relation. Moreover the flow time for a given liquid is not the same in the three viscometers; hence the conversion formulae or tables are employed to compare the results of such instruments. Tables of conversion are given in Appendix F.

4.9.1 *Redwood Viscometer*

The viscometer is named after its inventor Sir Boverton Redwood. The Institute of Petroleum recommended its use for oils for which the flow times, for 50 cm^3 of oil, lies in-between 2000s and 30 s under specified test conditions. The viscometer

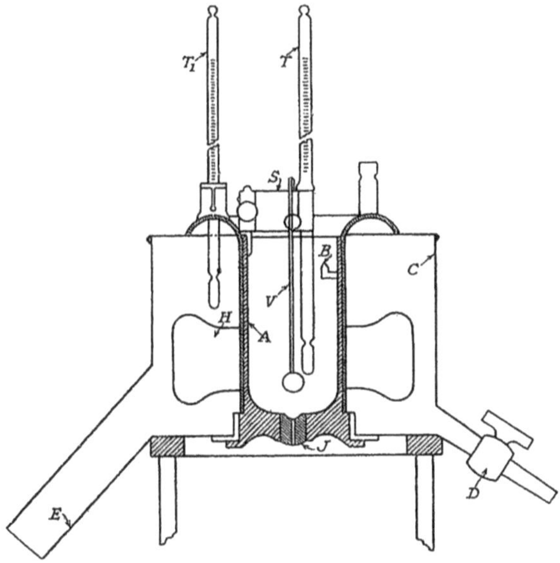

Fig. 4.9 Redwood viscometer

essentially consists of a flow cup and thermo-control system. Diagrammatically it is shown in Fig. 4.9.

4.9.1.1 Cup

The flow cup A is heavily silver plated to avoid corrosion and has an agate jet J(capillary) of specified dimensions. The jet is accurately bored and polished to the following dimensions:

Length of the jet 10 ± 0.05 mm

Minimum Internal diameter 1.62 mm

The pointer B in the cup indicates the level to which the oil under-test is to be filled. The valve V with a rod sits snugly on the opening of the jet J. This is used to start and stop of the oil flow.

4.9.1.2 Thermo-Control Bath

The cup is placed in a water bath C, which has the stirrer H having 4 vanes. It is used to stir water so that a uniform temperature is obtained. The side tube E is inlet for hot water and D with a tap is the outlet. Thermometers T and T_1 measure the temperature of the bath and the oil in the cup. To ensure the verticality of the cup there is a circular level which is sensitive to 0.2 degree of arc.

4.9.1.3 Working

The cup before use is washed with a suitable solvent like Benzene, petroleum spirit etc. and cleaned with some soft material such as tissue paper. Care is taken for cleaning the jet J, no fibrous material or any other thing should be left within the jet J. The oil under-test is heated separately within a few degrees of the test temperature. The oil is poured in the cup. The bath temperature is adjusted so that oil reaches the temperature of test uniformly through out its bulk. The level of the oil is adjusted to R, by moving the valve V vertically. The efflux time of the oil is then measured with a precision of 0.2 s by raising the valve V to fill a calibrated flask up to 50 ml mark.

The efflux time should be measured at the specified temperature. The specified temperature is attained by very slow heating or cooling the water bath. So that desired temperature reaches during midway of the measured efflux time. One measurement of efflux time is taken while the temperature is rising and second observation is taken while the temperature is falling. In each case rate of rise or fall of the temperature must be same and should be such that specified temperature is reached midway of the efflux time. The mean of the two efflux time is taken and the viscosity of the oil is reported in "Redwood seconds".

The more viscous oils sometimes show hysteresis and to eliminate its effect the oil is first heated to a specified temperature and is allowed to cool for 24 h before carrying out the viscosity test.

4.9.2 Redwood No 2 Viscometer

For more viscous oils, whose efflux time is more than 2000s in Redwood No 1 viscometer, the Redwood No 2 was adopted in such a way that efflux time for same oil in Redwood No 2 viscometer is exactly one tenth of that in Redwood No 1 viscometer.

The instrument is almost same as the Redwood No 1 viscometer except the dimensions of the jet in the cup and the method of attaching the bath.

4.9.2.1 Jet Dimensions

The jet dimensions are:

Internal length of jet J	50 $\pm$ 0.2 mm
External diameter	10.0 mm approximately
Minimum internal diameter	3.80 mm

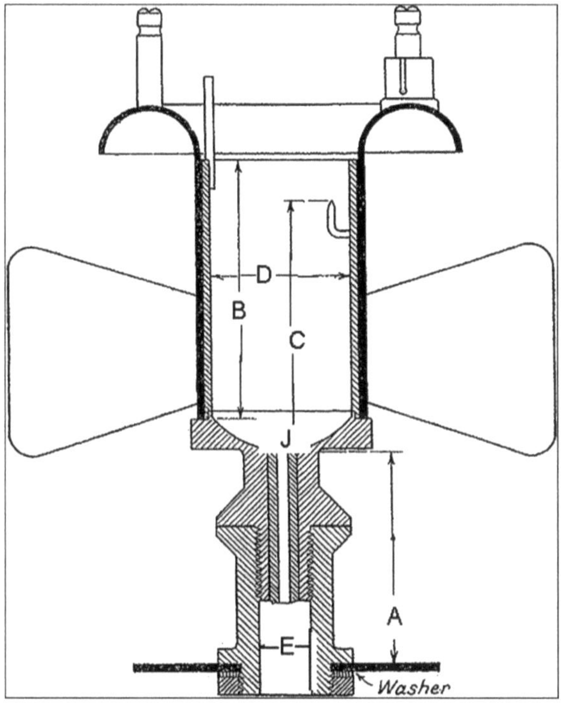

Fig. 4.10 Redwood No 2 viscometer

4.9.2.2 Construction Redwood No 2 Viscometer

In Redwood No 2 viscometer the method of attaching the bath gives better temperature control on the jet itself. Line diagram of Redwood No 2 viscometer is given in Fig. 4.10.

The cup has the diameter D = 46.5 mm and height B = 86 mm. The height, of the point from the top of the jet J to which the oil is to be filled, is C = 82.5 mm, i.e. initial pressure head is 82.5 mm. A—the distance between washer and the top of the jet is 70 mm and E is the internal diameter of the receiving flask.

The efflux time is measured in the same way as in Redwood No 1 viscometers, and viscosity is given in terms of Redwood seconds.

The results in each case are reported in seconds, stating which viscometer was used and the temperature of measurement. For example Viscosity Redwood No 1 at 140 °F is 350 s.

4.9.2.3 Conversion of Redwood Seconds in Absolute Units of Viscosity

Conversion of redwood second T_R into absolute units of measurements is basically carried out with Poiseuille's formula with kinetic energy term.

$$\nu = AT_R - \frac{B}{T_R} \tag{4.28}$$

Efflux time T_R is substituted by Redwood seconds. The values A and B for Redwood No 1 viscometer, calibrated by National Physical Laboratory UK are:

For T_R between 40 to 85 s	$A = 0.264$ and $B = 190$
For T_R between 85 to 2000 s	$A = 0.247$ and $B = 65$

Here ν is the kinematic viscosity in centistokes (cSt) or in mm^2/s.

Due to lack of exact standardization of the Redwood No. 2 viscometer, no formula for converting redwood seconds to absolute units of viscosity could be given. However Barr [14] suggested that A may be taken as 0.260 and B as 40 in the (4.28) for converting Redwood seconds to absolute units of viscosity.

4.9.3 Saybolt Universal Viscometer

Saybolt Viscometer is the American equivalent of British Redwood No 1 Viscometer. ASTM (American Society of Testing Materials) specified that Saybolt Viscometer should be used for lubricating oils and the efflux time for such oils should not be less than 32 s.

4.9.3.1 Construction of Saybolt Universal Viscometer

The construction of the viscometer is almost similar to that of Redwood viscometer. A narrower cup and bath extending well below the bottom of the outlet tube O are used so that better temperature uniformity of the oil under test is obtained. The viscometer is entirely made of a hard and corrosive resistant material, such as Stainless steel or Monel metal. The cup is shown in Fig. 4.11.

In this case the oil-level of the under-test oil is fixed by allowing the liquid to overflow. The over flown liquid is collected into the special rim B. The water bath extends up to the level KK and level of water LL is kept slightly above the top of the cup. A cork C fitting into a recess below the outlet tube O is used to start the flow. The oil is collected in a calibrated flask of 60 cm^3. The dimensions of the outlet tube O are:

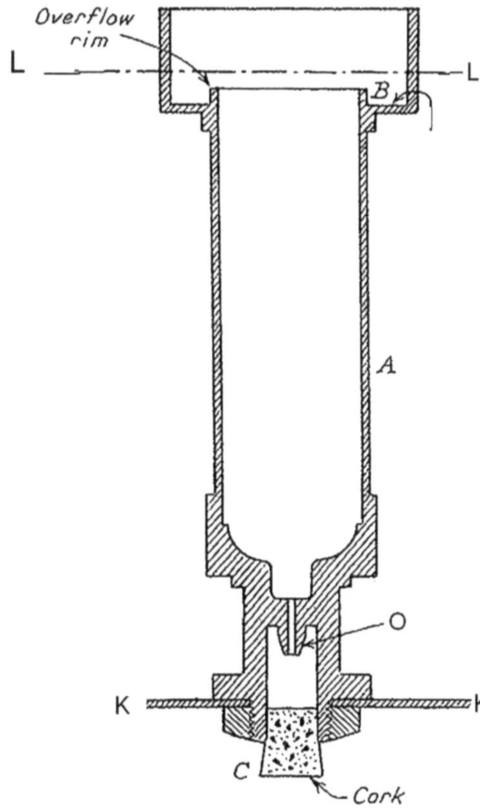

Fig. 4.11 Cup for saybolt universal viscometer

Inside diameter of the outlet tube	1.765 ± 0.015 mm
Outside diameter of the outlet tube at lower end	3.0 ± 0.2 mm
Length of the outlet tube	12.25 ± 0.10 mm

The efflux time is measured by delivery of 60 cm^3 of oil. Proper attention is given that repeated observations of efflux time are within 1 %.

4.9.3.2 Conversion of Saybolt Seconds in Absolute Units of Viscosity

In this case also formula for conversion is similar to that used for Redwood seconds Three sets of constants are given depending up on the efflux time. The equation used is

$$v = AT_{SU} - \frac{B}{T_{SU}} \tag{4.29}$$

Here v is the kinematic viscosity in centistokes (cSt) or in mm^2/s and T_{SU} represent the Saybolt seconds

The constants are

(i) For T_{SU} less than 100 s	A = 0.226 B = 195
(ii) For T_{SU} more than 100 s	A = 0.220 B = 130
(iii) For T_{SU} more than 200 s	A = 0.220 B = 0.00

4.9.4 *Saybolt Furol Viscometer*

It corresponds to British Redwood No2 viscometer and meant for viscous oils i.e. for those oils for efflux time in a Saybolt Universal Viscometer is inconveniently large. The Saybolt Furol viscometer is identical to the Saybolt universal viscometer except its outlet.

The dimensions of the outlet tube are:

Inside diameter of the outlet tube	3.15 ± 0.02 mm
Outside diameter of the outlet tube at the lower end	4.3 ± 0.3 mm
Length of the outlet tube	12.25 ± 0.1 mm

The results of measurement are obtained in seconds (s) and symbolized by T_F.

Similar to Saybolt Universal viscometer an accuracy of ±1 % is claimed for this viscometer also.

4.9.4.1 Conversion of Kinematic Viscosity *in SI to* T_{SU}

The use of given values of A and B to convert Redwood or Saybolt seconds in kinematic viscosity from (4.27) or (4.28) give only approximate values of kinematic viscosity. Recently ASTM has issued a standard D 2161-10, in which T_{SU} (Saybolt universal seconds) have been obtained for different values of kinematic viscosity v in $mm^2\ s^{-1}$or cSt. The tables cover a range from almost 1 to 500 $mm^2\ s^{-1}$. The corresponding values of T_{SU} in seconds (s) have been given for temperatures of 100 °F and 210 °F.

The tables start from 1.77 $mm^2\ s^{-1}$ for T_{SU} at 210 °F and from 1.81 $mm^2\ s^{-1}$ for T_{SU} at 100 °F. Tables give values of T_{SU} in terms of kinematic viscosity in $mm^2\ s^{-1}$ in steps of

$0.01\ mm^2\ s^{-1}$ up to $10\ mm^2\ s^{-1}$
$0.02\ mm^2\ s^{-1}$ from 10 to $20\ mm^2\ s^{-1}$
$0.05\ mm^2\ s^{-1}$ from 20 to $50\ mm^2\ s^{-1}$
$0.1\ mm^2\ s^{-1}$ from 50 to $100\ mm^2\ s^{-1}$
$0.2\ mm^2\ s^{-1}$ from 100 to $200\ mm^2\ s^{-1}$
$0.5\ mm^2\ s^{-1}$ from 200 to $500\ mm^2\ s^{-1}$

Similarly tables for conversion of kinematic viscosity to Furol seconds T_F have given in steps of:

$1\ mm^2\ s^{-1}$ from 48 to $900\ mm^2\ s^{-1}$
$2\ mm^2\ s^{-1}$ from 900 to $1300\ mm^2\ s^{-1}$

It appears that ASTM collected large amount of data about the kinematic viscosity and T_{SU} and T_F for various oils and fitted a smooth curve to the original experimental data. The derived equations are given as follows:

$$T_{SU} \text{ at } 100\,^{\circ}\mathrm{F} = 4.6324\nu + \frac{1.0 + 0.0326\nu}{(3930.2 + 262.7\nu + 23.97\nu^2 + 1.646\nu^3)10^{-5}}$$

$$T_{SU} \text{ at } t\,^{\circ}\mathrm{F} = T_{SU} \text{ at } 100\,^{\circ}\mathrm{F}\,\{1.0 + 0.00006(t - 100)\}$$

$$T_F \text{ at } 122\,^{\circ}\mathrm{F} = 0.4717\nu + \frac{13924}{\nu^2 - 72.59\nu + 6816}$$

$$T_F \text{ at } 210\,^{\circ}\mathrm{F} = 0.4792\nu + \frac{5610}{\nu^2 + 2130}$$

4.9.5 Engler Viscometer

Like other viscometers discussed in this section, this viscometer is also named after its inventor Engler. This is one of the oldest viscometer as it was officially recognized by German Railways as early as 1894. It consists of a cup and a jet. The specifications for jet and the cup were prepared by German standardization agency and other two concerned organizations in 1907 and slightly modified in 1928.

The cup is outlined in Fig. 4.12, is gold plated from inside and the jet is made either of platinum or its inside is coated with thick layer of platinum.

The values of symbols indicating different dimensions, in mm, of the cup are indicated in the table below:

Dimensions	B	C	D	E	F	G	H	K
Values	106	25	52	2.9	2.8	4.5	20	3

Fig. 4.12 Engler viscometer

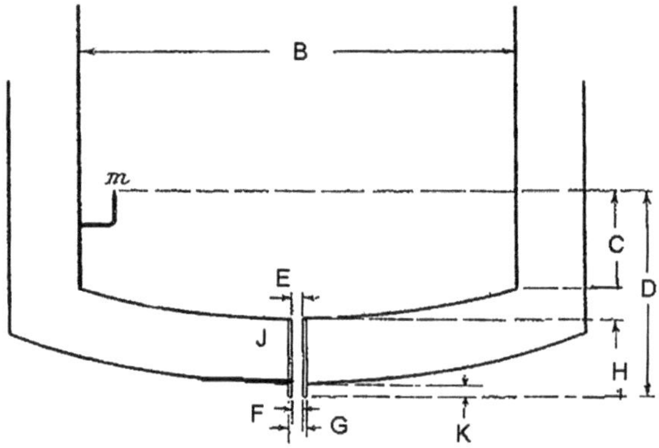

It may be noticed that the jet J is slightly tapered (2.9 mm at the top and 2.8 mm at the lower end).

The specified dimensions of the jet with permissible tolerances are

Length	20.0 ± 0.10 mm
Internal diameter of the jet at top	2.9 ± 0.02 mm
Internal diameter at the lower end	2.8 ± 0.02 mm

In order to start or stop the flow, a stopper is placed in the top of the jet. Three points are equally fixed along its circumference to fix the level of the liquid and are also used for leveling the cup.

The time of flow for 200 cm^3 of the oil is measured in seconds. Viscosity is expressed as the ratio of the efflux time of oil and for same volume of water at 20 °C and is termed as Engler degree. Efflux time for water at 20 °C is also called water value.

Main source of error is the temperature gradient in the oil. This is because the area of cross-section is relatively large to its depth.

For conversion of Engler degrees E into absolute units of viscosity, Ubbelohode issued a conversion formula

$$\frac{\eta}{\rho} = 0.073E - 0.0631/E \tag{4.30}$$

Erk [15] expressed kinematic viscosity $\nu = \eta/\rho$ in exponential way. His formula gives a better approximation.

$$\frac{\eta}{\rho} = E \times (0.0760)^{(1-1/E)} \tag{4.31}$$

4.10 Bubble Viscometer

Bubble viscometers are used to quickly determine kinematic viscosity of some liquids such as resins and varnishes. The time required for an air bubble to rise is directly proportional to the viscosity of the liquid, so the faster the bubble rises, the lower the viscosity. The Alphabetical Comparison Method uses 4 sets of lettered reference tubes, A5 through Z10, of known viscosity to cover a viscosity range from 0.005 to 100,000 mm^2/s. The Direct Time Method uses a single 3-line times tube for determining the "bubble seconds", which may then be converted to stokes (1 centistoke = 1 mm^2/s).

References

1. Couette MM (1890) Ann Chim et Phys 21:433
2. Merrington AC (1949) Viscosmetry. Edward Arnold & Co, London, pp 30–42
3. Ungar (1934) Koll. Zietes, 69:30, 164
4. Ungar (1935) Koll. Zietes 70:227
5. Ungar (1935) Koll. Zietes, 71:16
6. Schlichting H, (1955) Boundary layer Theory, Trans. By Kestin J, McGraw Hill Book Co, pp 75–80, 443–447
7. Maxwell JC (1866) Phil Trans Roy Soc (London) 156:249
8. Hubbard RM, Brown GG (1943) Ind Eng Chem 35:1276
9. Hersay MD (1916) Theory of torsion and rolling ball viscometers, their use in measuring the effect of pressure on viscosity. J Wash Acad Sci 6:525–530
10. Flowers AE (1914) Viscosity measurement and a new viscometer. Proc Am Soc Test Mater 14:565–616
11. Forgacs RL (1965) Improved Torsion viscometer for Polymer studies. Rev Sci Instrum 36:307–314
12. http://en.wikipedia.org/wiki/Viscometer#Micro-Slit_Viscometers
13. Walters K (1975) Rheometry. Chapman and Hall, London
14. Barr G (1931) Viscometry. Oxford University Press, Oxford. p 94
15. Erk S (1927) Forschungsarbeiten V. D. I. p 288

Chapter 5
Oscillating Viscometers

Abstract As per the heading of the chapter, it contains oscillating type viscometers. Theory of damped vibrations and various methods of determining the time period and logarithmic decrement have been briefly described. Expressions, implicitly containing viscosity of the liquid in which cylindrical and spherical bodies are oscillating, are given. The importance of oscillation viscometer may be gauged from the fact that these are the only ones other than capillary viscometers which are used for determination of viscosity of water. Viscometers used by Roscoe and Bainbridge, Torklep and Oye, Kestin and Shankland and Berstad et al. have been sketched. NBS torsion viscometer and its theory have also been briefly described.

5.1 Oscillating Viscometers

A body (cylinder hollow or solid, solid disc, or sphere) is suspended through a suspension wire from the rigid support making it as a torsion pendulum. If the pendulum is in vacuum and is made to oscillate once in horizontal plane with torsion wire as axis, then it will continue to oscillate indefinitely. However if the same pendulum is made to oscillate in a fluid, its amplitude will go on decreasing and eventually will stop. During the motion its time period will remain the same though it will be slightly more than that in vacuum. The reason for decreasing in amplitude is obviously the viscous forces exerted by the fluid which will be some function of the geometry of the oscillating body and the viscosity of the surrounding fluid. Such an arrangement is called an absolute oscillating viscometer. The word absolute is used as it gives the viscosity of the liquid in terms of dimensions of oscillating body, measured time period and decrement in amplitude without reference to the viscosity of any other liquid.

Detailed theory of oscillating viscometers having right circular cylinder, cup and sphere as oscillating body, have been worked out by Prof Newell Group at Brown University [1–6].

S. V. Gupta, *Viscometry for Liquids*, Springer Series in Materials Science 194,
DOI: 10.1007/978-3-319-04858-1_5,

5.2 Damped Oscillations

If a body is suspended through a suspension wire from the rigid support and the body is made to oscillate then the suspension wire provides the restoring torque. The system is made to oscillate in vacuum it will continue to oscillate with constant amplitude say with a time period T_0. However when the system is made to oscillate in a fluid a drag force acts on the rotating body, which is a function of viscosity of the fluid and geometry of the oscillating system. The drag force affects in two ways (1) the change in its time period (2) the decrease in its amplitude after every half cycle. Differential equation applicable for such a motion is that of a damped oscillator and is given by:

$$\left(I \cdot D^2 + r \cdot D + k\right)\theta = 0 \tag{5.1}$$

where

I is the moment of inertia of the oscillating body
k is the restoring moment of suspension and
r is a function of various parameters one of which is viscosity of the fluid.

If $\theta = \theta_0 \exp(mt)$ be a solution of (5.1) then on substitution the value of θ in (5.1), we get

$$Im^2 + rm + k = 0$$

Giving

$$m = \frac{-r \pm \sqrt{r^2 - 4Ik}}{2I} \tag{5.2}$$

Hence general solution of (5.1) is given as

$$\theta = \theta_0 \exp\left(\frac{-r \pm \sqrt{r^2 - 4Ik}}{2I} \cdot t\right) \tag{5.3}$$

If $r^2 - 4Ik$ is positive then θ will be a continuously decreasing function of time with no oscillatory motion, hence for oscillatory motion it is necessary that $(r^2 - 4Ik)$ is negative.

Let us write

$$\frac{\sqrt{r^2 - 4Ik}}{2I} = \omega \mathrm{i}$$

When time is measured from equilibrium position, (5.3) is written as

$$\theta = \theta_0 \left\{\exp\left(\frac{-rt}{2I}\right)\right\} \sin(\omega t) \tag{5.4}$$

In (5.4), ω is the angular frequency of the oscillatory motion and is equal to $\frac{2\pi}{T}$. Here T is the time period of the oscillatory motion. If δ is the logarithmic decrement in amplitude, then δ can be expressed as

$$\frac{2\pi}{T}\delta = \frac{r}{2I},$$

Giving (5.4) as

$$\theta = \theta_0 \exp\left(-\frac{2\pi\delta}{T}t\right)\sin\left(\frac{2\pi}{T}t\right) \tag{5.5}$$

Here the motion is oscillatory with fixed time period T but its magnitude is decreasing. Hence the motion is defined by two quantities namely time period T and the logarithmic decrement δ. Viscosity of the fluid surrounding the oscillating pendulum is implicitly related with δ and T, hence it is imperative to discuss the measurement of δ and T.

5.3 Measurement of δ and T

5.3.1 Logarithmic Decrement by Linear Measurement

The logarithmic decrement δ is determined by measuring the amplitudes of consecutive vibrations. Time period T is determined simultaneously by measuring the time taken for the certain number of oscillations. About 50 consecutive amplitudes and time taken between each turning point are measured.

$$\text{Let } t_{\mathrm{n}} = (2n-1)T/4$$

Then

$$\begin{aligned} \theta_n &= \theta_0 \exp\left\{\frac{2\pi\delta}{T}\frac{(2n-1)T}{4}\right\}\sin\left\{\frac{2\pi}{T}\frac{(2n-1)T}{4}\right\} \\ \theta_n &= \theta_0 \exp\left\{-\frac{(2n-1)\pi\delta}{2}\right\}(-1)^n \end{aligned} \tag{5.6}$$

Here n is any positive integer including zero.

The angular deflection θ of the oscillating body is converted into a linear displacement A by fixing a mirror to it and measuring its displacement on a linear scale placed at a distance d from the plane of the mirror. Giving us

$$A = 2d\theta$$

Giving (5.5) as

$$A = A_0 \exp\left(-\frac{2\pi\delta}{T}t\right)\sin\left(\frac{2\pi}{T}t\right) \tag{5.7}$$

In practice measurement of time may not start exactly at the time when displacement is zero. Let the real zero (equilibrium position) is at a distance a from the arbitrary zero. In that case, if $b_1, b_2,$ $b_3,$ $b_4, b_5 \ldots b_n$ be the amplitudes measured from some arbitrary zero. and $A_1, A_2, A_3, A_4, A_5 \ldots A_n$ be the real amplitude at time t equal to $T/4, 3T/4, 5T/4, 7T4\ldots\ldots(2n+1)T/2$ respectively. Then referring to Fig. 5.1, we get

$$\begin{array}{llll} A_1 = b_1 + a & & b_1 - b_3 = A_1 - A_3 \\ A_2 = b_2 - a & & b_2 - b_4 = A_2 - A_4 \\ A_3 = b_3 + a & \text{Giving} & b_3 - b_5 = A_3 - A_5 \\ A_4 = b_4 - a & & \\ A_5 = b_5 + a & & \\ A_6 = b_6 - a & & \end{array} \tag{5.8}$$

Writing the value of θ's from (5.27) and substituting it in (5.28), we get

$$\begin{aligned} b_1 - b_3 &= A_0 \exp(-\pi\delta/2)[1 - \exp(-2\pi\delta)] \\ b_2 - b_4 &= A_0 \exp(-3\pi\delta/2)[1 - \exp(-2\pi\delta)] \\ b_3 - b_5 &= A_0 \exp(-5\pi\delta/2)[1 - \exp(-2\pi\delta)] \\ b_4 - b_6 &= A_0 \exp(-9\pi\delta)/2[1 - \exp(-2\pi\delta)] \\ &\text{etc.} \end{aligned} \tag{5.9}$$

From (5.9) A_0 and δ can be determined. T is determined by measuring the time of say 50 oscillations. Time period and amplitude measurements can be done electronically with a counter and linear photo cell or manually with employing lamp and scale arrangement.

Roscoe and Bainbridge [7] used lamp and scale arrangement and measured δ and T manually.

5.3.2 Logarithmic Decrement by Time Measurement

Kestine and Khalifa [8] expressed logarithmic decrement in terms of time intervals elapsed between crossing the zero line and a line of fixed amplitude. We see in Fig. 5.1 that time interval τ increases with the subsequent vibrations. Let us consider mth and nth vibrations, the total time t_m and t_n taken to attain the same amplitude are given as

$$t_m = mT + \tau_m \text{ and } t_n = nT + \tau_n \tag{5.10}$$

Fig. 5.1 Damped motion

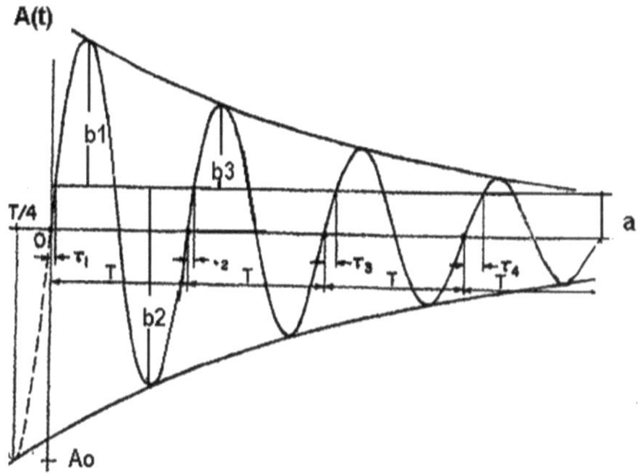

Substituting these values of time in (5.5), we get

$$a = \exp\left\{-\frac{2\pi}{T}\delta(\tau_m + mT)\right\}\sin\left\{\frac{2\pi}{T}(mT + \tau_m)\right\} \tag{5.11}$$

$$a = \exp\left\{-\frac{2\pi}{T}\delta(\tau_n + nT)\right\}\sin\left\{\frac{2\pi}{T}(nT + \tau_n)\right\} \tag{5.12}$$

Dividing (5.11) by (5.12) and taking the logarithm, we get

$$-(2\pi/T)\delta(\tau_m - \tau_n + mT - nT) = \log\left[\frac{\sin(2\pi\tau_n/T)}{\sin(2\pi\tau_m/T)}\right] \tag{5.13}$$

Giving

$$2\pi\delta = \frac{\log\left[\frac{\sin(2\pi\tau_n/T)}{\sin(2\pi\tau_m/T)}\right]}{(n-m) + (\tau_n - \tau_m)/T} \tag{5.14}$$

Experimental arrangement used by Kestin and Khalifa [8] is indicted in Fig. 5.2. The numerals in the figure represent the following:

1 is a 1 mW He Ne laser, 2 is the front face coated mirror fixed to the oscillating system, 3 and 4 are the slits containing the photo-resistors placed on the 0–0 line at a distance of 2 m from the oscillating mirror 2. Slit 3 is mounted on a micrometer carriage and slit 4 is fixed but the two slits are in the same vertical plane, which is parallel to the vertical plane of the mirror, when it is equilibrium position. 5 and 6 are two diodes which send the current to the digital time counters 9 and 10 through the trigger amplifiers 7 and 8. Each of counter works on 50 MHz. Counter 9 measures the time period T and Counter 10 gives the values of τ.

The slit containing the photo-resistor is so adjusted that it intercepts the reflected light from the mirror 2 when it is stationary, i.e. in equilibrium position of the oscillating system, hence the zero amplitude is defined. Photo-resistor 4 is

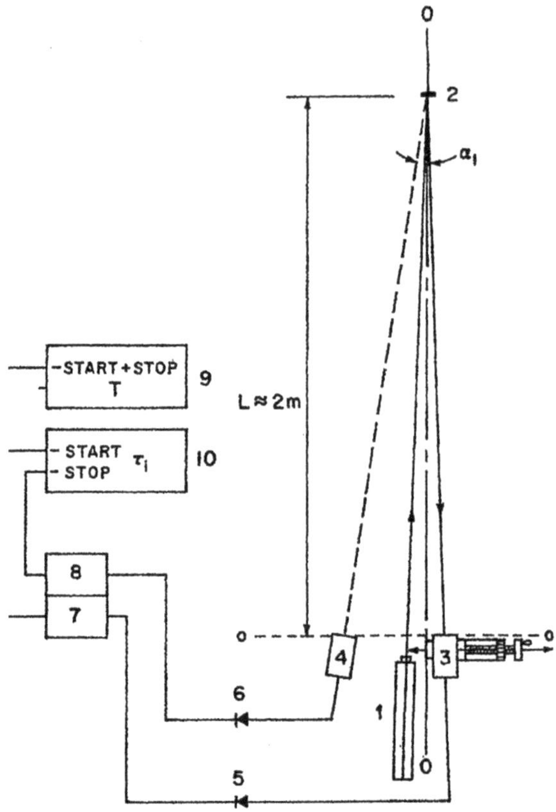

Fig. 5.2 Measurement of time period and time intervals τ_m

capable to intercept the reflected light and thus defines the given fixed amplitude a. The impulse which starts counter 9 also starts counter 10, but counter 10 is stopped from the signal from photo-resistor 4 through trigger amplifier 8.

5.3.3 Logarithmic Decrement by Time Measurement Between Two Fixed Points

As the amplitude of forced vibration is continuously decreasing with constant time period, time taken to traverse a distance between two fixed points in each subsequent excursion will be increasing. Using the method cited in Sect. 5.3.2, instead of taking zero line crossing as one fixed point and other at any other amplitude, one may take any other two fixed points. The time intervals are measured for at least three successive intervals, from which all the three unknowns namely θ_0, δ and T can be evaluated. As the δ and T are some complex functions of viscosity η and dimensions of the oscillating body, viscosity is calculated by numerical methods.

5.4 Viscosity Equations

Implicit equations containing viscosity η depend upon the shape and dimensions of the oscillating body. Having measured δ, T and other dimensions of the oscillating body, the value of viscosity η is calculated by successive approximations to the required degree of precision.

5.4.1 Right Circular Cylinder as Oscillating Body

For the right circular cylinder of radius R, height $2h$, dynamic viscosity η is calculated from either of the two equations [9]:

$$\left(\pi\rho hR^4/I\right)\left(A(p-q\delta)/x+\ \mathrm{B}/x^2+\mathrm{Cq}/\mathrm{x}^3\right)\ =2(\delta-\delta_0/\omega) \tag{5.15}$$

$$\left(\pi\rho hR^4/I\right)\left(A\left(p\delta+q\right)/x+B\delta/x^2+Cp/x^3\right)=1/\omega^2-1+\left(\delta-\delta_0/w\right)^2 \tag{5.16}$$

Here

δ_0 is the logarithmic decrement in air

$$A=4+R/h$$

$$B=\frac{16(4\pi-3\sqrt{3})}{9\pi\sqrt{3}}(R/h)+6=2.407949R/h+6 \tag{5.17}$$

$$\mathrm{C}=(17/9)(R/h)+3/2$$

$$\omega=T_0/T \tag{5.18}$$

$$p=\frac{1}{\sqrt{(2\left\{\delta+(1+\delta^2)^{1/2}\right\}}}$$

$$q=1/2p \text{ and } x=R(2\pi\rho/\eta T)^{1/2}$$

T_0 and T are the time periods in air and liquid respectively and ρ is the density of the liquid.

5.4.2 Sphere as an Oscillating Body

The working equations for sphere of diameter $D=2R$ as given by Verschaffelt [10] are as follow:

$$\eta = \frac{3DI\{1 + 1/\omega^2(1 + D^2)\}}{4TR^3(2 + b_1R + p_1)} \tag{5.19}$$

$$\eta^{1/2} = \frac{3I\{-1 + 1/\omega^2(1 + D^2)\}}{4TR^4\gamma(1 - q_1)\rho^{1/2}} \tag{5.20}$$

Here

$$\begin{aligned}
p_1 &= (b_1R + 1)q_1 \\
q_1 &= \frac{1}{\left\{(b_1R + 1)^2 + b_2^2R^2\right\}} \\
b_1 &= \{\pi\rho/\eta T\}^{1/2}\left\{-D + (1 + D^2)^{1/2}\right\}^{1/2} \\
b_2 &= \{\pi\rho/\eta T\}^{1/2}\left\{D + (1 + D^2)^{1/2}\right\}^{1/2} \\
\gamma &= \{\pi/T\}^{1/2}\left\{D + (1 + D^2)^{1/2}\right\}^{1/2} \\
D &= \delta - \delta_0
\end{aligned} \tag{5.21}$$

5.5 Viscometer Used by Roscoe and Bainbridge

5.5.1 Viscometer

The viscometer used by Roscoe and Bainbridge [7] is outlined in Fig. 5.3. The rotatable suspension head was similar to that used by Andrade and Chiong [11, 12]. The bifilar suspension S consisted of a loop of tungsten wire of diameter 0.14 mm and length 680 mm. The length is the distance from the suspension head to the wheel W. The frame F holding this wheel and the mirror M, the disc D supporting the inertia rings R and the threaded tube T all formed one unit which was carefully machined to secure alignment. The thread on the brass collar C fixed to the stem of the bulb fitted loosely into the threaded tube, so constant alignment of the bulb was secured by contact between the horizontal surfaces of the base of the tube and the projection below the threaded part of the collar. The suspension wire was extended on the frame above the wheel after the water filled bulb had been screwed into position.

The rotation of the bulb was observed by means of the image of a fine thread formed by lens and reflected by the plane mirror M on to a suitably curved scale placed 3 m away.

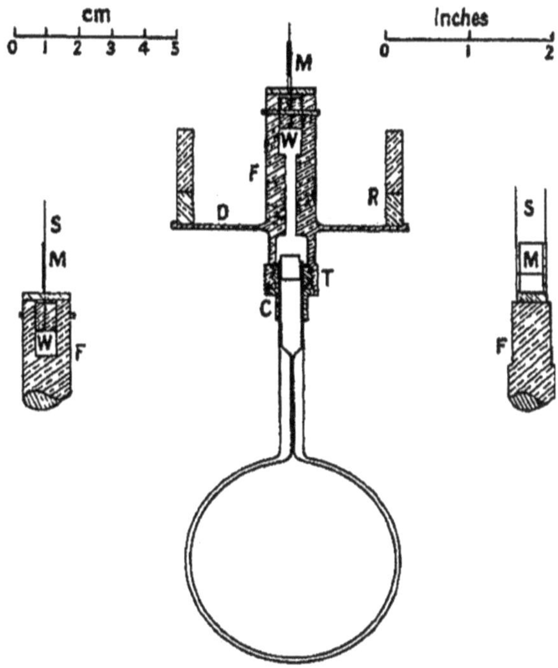

Fig. 5.3 Viscometer (Roscoe and Bainbridge)

Basic equation used by Roscoe and Bainbridge [7] was:

$$I\delta = \frac{2}{3}R^4(\pi\eta\rho T)^{1/2}F(\delta,p) \tag{5.22}$$

Here I is the moment of inertia of the suspended part, T is the time period, δ is the logarithmic decrement, R is the radius of the bulb, η is the viscosity and ρ is the density of the liquid. The symbol p stands for $(\pi\rho/\eta T)^{1/2}$. F(δ,p) is a function of δ and p. F approaches 1 for small values of δ and very large values of p.

To calculate viscosity, the actual temperature of measurement, logarithmic decrement δ and time period T are measured.

5.6 Viscometer Used by Torklep and Oye

Principle: Torklep and Oye [13] measured the damping of a right circular cylinder (disc and cup) oscillating in contact with the liquid under test. Logarithmic decrement is determined by digital method solely based on measurements. Dimensions and moment of inertia of the oscillating cylinder are measured to best of accuracy available. The equation (5.15) is used to calculate viscosity η of the liquid in contact with the oscillating system.

5.6.1 Support System

A 100 kg iron base plate resting on four adjustable vibration absorbing supports is placed on a concrete block cast on a rubber mat. The viscometer frame stands on the iron base plate. A bracket fastened to the top of the stand supports the viscometer. The stand is capable for accommodating, if necessary, a viscometer furnace which can be adjusted laterally and vertically.

The Fig. 5.4 gives a cross-sectional view of the viscometer without the stand. A precisely ground central plate K screwed on the above mentioned bracket supports all the mechanical components that surround the torsion wire and the pendulum. Above the plate there is a temperature controlled braking unit G and a double walled stainless steel tube C with wire suspension head A, B; a unit bellow the plate comprises a vacuum bellows connection M to the furnace T; a temperature controlled cold-finger Q and radiation shields. The space around the viscometer can be evacuated to about 10^{-3} Pa for drying samples and change of atmosphere or for conducting experiments in vacuum.

5.6.2 Torsion Pendulum

The length of the torsion pendulum R from the torsion wire grip to the centre of the oscillating body is 560 mm. The upper part is made of stainless steel, tapered in steps for a maximum diameter of 12 mm as shown in Fig. 5.3. Permanently joined to this part is 3 mm diameter Pt Ir (Pt 20 %) rod, terminating in a small flange and a thread with central pin. The detachable oscillating body, which is equipped with a matching flange, is screwed on to the pendulum after this has been installed in the viscometer. A Pt Ir (20 % Pt) solid cylinder is preferably used as the oscillating body.

The pendulum and the solid cylinder are checked for straightness with the help of lathe and microscope. Special attention is paid to the flanges. The whole oscillating system is thus set to a linearity of 0.02 to 0.03 mm or better.

The dimensions of the cylinder (the oscillating body) are measured within ± 1 μm. Roundness of the cylinder is checked at a magnification of 10,000 at different sections of the cylinder. Deviation from the circularity and radius of the cylinder at different sections should be within 1 μm.

5.6.3 Torsion Wire

The torsion wires are of 8 % Pt and 580–600 mm long. A wire of 0.3 mm diameter is good for a solid cylinder of 350 g, for heavier pendulum of say 750 g a wire of diameter 0.5 mm should be taken. The suspension wire has low internal friction

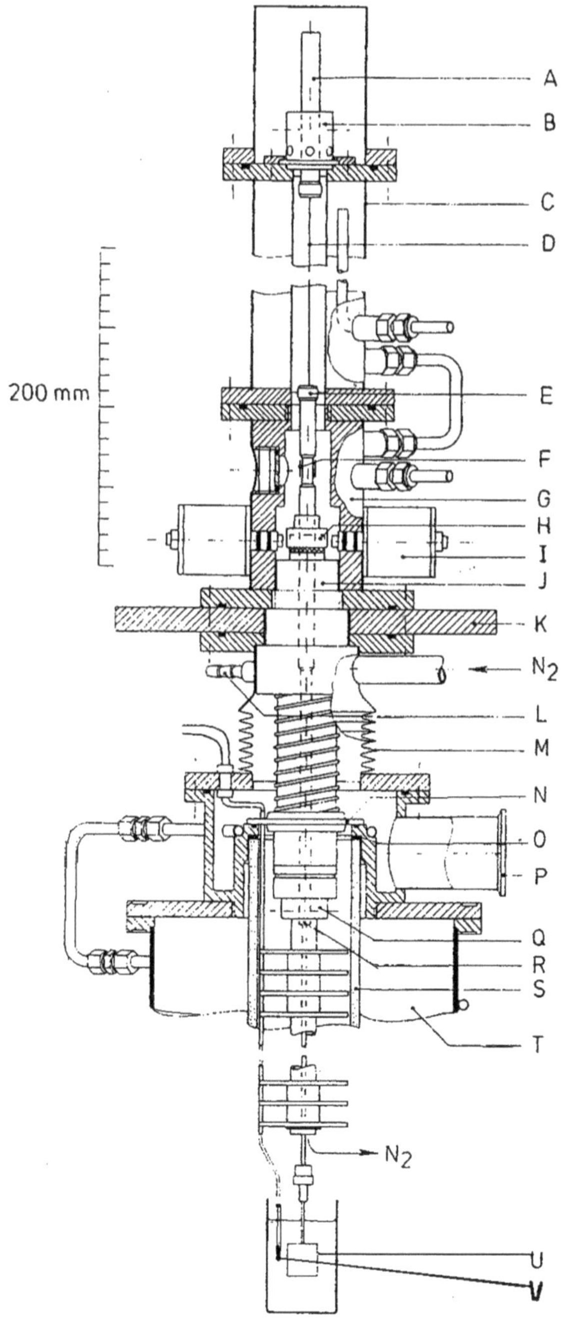

Fig. 5.4 Cross-sectional view of viscometer. *A* vertically adjustable bolt, *B* rotatable bolt suspension, *C* double walled tube supporting and thermo stating the torsion wire, *D* torsion wire, *E* torsion wire grip, *F* Platinised mirror, *G* Temperature controlled housing, *H* cup shaped aluminium rotor, *I* six electromagnets only 2 visible, *J* toroid supporter, *K* central plate, *L* inlets for thermo stated water to cold fingers, *M* vacuum bellows, *N* spring loaded ring seal, *O* furnace tube vacuum seal, *P* connection pump station, *Q* lower end of thermo stated cold-finger, *R* torsion pendulum, *S* Furnace tube 600 mm long, *T* furnace shown in raised operation, *U* Pt Ir (20 % Pt) cylinder, *V* Pt/Pt 10 % Rh thermocouple

and high torsion constant stability and is annealed at 1000–1100 °C. If some accidental bending occurs the suspension should be annealed in situ.

The torsion wire suspension head A, B consists of a wire grip and precisely guided steel bolt locked by three screws at 120° each. The suspension bolt, torsion wire and pendulum are removed as a single unit by lowering it through the viscometer. The suspension system is covered with a vacuum tight cup. The wire grip in principle is similar to chucks used in precision lathe.

The heating element is a bifilar resistance wire and is wound continuously over the most of the length of the furnace T.

5.6.4 Cross-Sectional View of the Viscometer

Figure 5.4.

5.6.5 Oscillation Initiator

The oscillation initiator is a torque motor comprising six stator coils I surrounding a thin-walled aluminium cup or rotor H fastened to the pendulum. Inside this cup is a stationary, low-magnetic-remanence iron toroid. The coils, which are matched on the basis of magnetic field measurements, have cores of the same material. All parts in the vicinity of the torque motor are made of non-magnetic materials.

The stator coils are connected to triacs that fire and switch off in the same sequence as zero of the mains voltage. The start pulse is pre-selective with effective duration varying from about 13–600 ms in steps of 20 ms. A typical start lasts 53 ms when using the Pt-Ir cylinder (Pt 90 %), fine control on initial amplitudes is provided by resistors in series with the coils. The excitation system is also used to bring the pendulum to rest in between by applying a direct current to the coils. Both, the braking period and the cut-off times are adjustable and pre-selected.

5.6.6 Measurement of δ and T

Logarithmic damping δ and time period T in water and air are measured by measuring the time interval required to travel the distance between the two fixed diodes for several consecutive excursions of the oscillating system. The two diodes are kept 100 mm apart and roughly symmetrical to the oscillation zero of the reflected beam. Laser beam sent through a lens system of one convex and two cylindrical lenses to form a line- shaped image is reflected from the mirror on the pendulum to fall upon two photodiodes. The distance between the planes of photodiodes and pendulum is kept 2 m.

The viscometer was so designed that the response of the damping to the viscosity was optimum.

5.6.7 *Calculation of Viscosity*

Having known δ, T, dimensions and moment of inertia of the oscillating cylinder, the viscosity η of the liquid is calculated by using (5.15).

5.7 Viscometer Used by Kestin and Shankland

In earlier viscometers [14], the disc used to oscillate between two fixed plates to counteract the effects of secondary flows. But to avoid errors in compensating for the differential thermal expansion, Kestin and Shankland [15] developed a viscometer in which there were no fixed plates. They also developed the necessary mathematical expressions to calculate viscosity by measuring damping constant and time period of the free oscillating disc.

Their working equations are as follows:

$$\frac{2}{\omega}\left(\frac{\delta}{\omega}-\delta_0\right)=g_0\rho y\left[g_1H_1+g_2y/\omega+g_3y^2/2g_4\omega+yD_R(\eta_0)/R\omega\right] \tag{5.23}$$

and

$$\left(\frac{\delta}{\omega}-\delta_0\right)^2+1-1/\omega^2=g_0\rho y\left[g_1H_2+g_2y\delta/\omega-g_3y^2g_4-yQ_I(\eta_0)/R\omega\right] \tag{5.24}$$

Here

$$y=\left(\frac{\eta}{\rho}\frac{T_0}{2\pi}\right)^{1/2},\qquad \omega=T/T_0,\qquad g_0=\pi R^4/I$$

$$g_1=1+2d/R,\quad g_2=\left[\frac{16}{3\pi}\left(\frac{4\pi}{3\sqrt{3}}-1\right)+\frac{3d}{R}\right]/R,\quad g_3=\left(\frac{17}{9}+\frac{3d}{4R}\right)/R^2,$$

$$g_4=\left[\frac{(\delta^2+1)^{1/2}-\delta}{2\omega}\right]^{1/2},\quad H_1=\frac{3g_4}{2\omega}-\frac{1}{8g_4^3\omega^3}\text{ and }H_2=\frac{3}{4g_4\omega^2}-g_4^3 \tag{5.25}$$

$Q_R(\eta_0)$ and $Q_I(\eta_0)$ denote the real and imaginary parts of $Q(\eta_0 i^{1/2})$ respectively.

In fact (5.15) and (5.16), given by Berstad et al. [5], are the modified forms of (5.23) and (5.24).

5.7.1 Original Viscometer due to Kestin et al.

The vertical section of the viscometer [15] is shown in Fig. 5.5. The material used for the viscometer body was Hastelloy C276. The lower part 3 is solid and upper part 7 is hollow containing the oscillatory part. The lower part is mounted on a large diameter self centering bearings 1 whose outer race is recessed in a thick base plate made of stainless steel. The base plate is maintained in a horizontal position by means of three levelling screws rest on another thick stain less steel plate. The plate itself is placed on a massive, vibration-free concrete pillar. The upper part 7 is attached to the lower part 3 by means of a large buttress-threaded stainless steel cap 8 and 10 bolts labelled as 12 pressing against a pressure ring 13. For very high pressure of 35 MPa extra malleable gold wire is used in conjunction O rings.

The PRT thermometers 16 come through the well sealed holes in the solid lower part 3.

The oscillating-disk 5 is made of Hastelloy C276 which oscillates between two Hastelloy fixed plates 4 and 6 made of same material (Hastelloy). The suspension wire 9 is made of a stress relieved 92 Pt-SW thin strand. The suspension wire is gripped by means of a threaded cap nut and fixed to a stub shank rigidly attached to the disk.

The upper end of the wire is similarly gripped at the top of the suspension cartridge. The top part of the suspension assembly is mounted on a precision-ground alumina column 10. The alumina column 10 is open over most of its height to provide access to the strand and is attached at its bottom end to the suspension mounting plate 14. The suspension mounting plate carries three vertical pillars along which the top fixed plate can slide vertically. The bottom fixed plate 4 is separated from the top fixed plate 6 by means of three spacers located on the three vertical pillars. The mounting plate 14 is attached to a thick-walled cylinder 15 which is concentric with the axis of rotation of the viscometer and is provided with two side windows to provide access to the disk and the fixed plates.

All the mating parts of the suspension cartridge are held concentric with one another and with the viscometer axis by means of fitting centering surfaces, the maximum admissible eccentricity being less than one strand diameter (0.1 mm). The top suspension assembly 11 is designed to permit controlled adjustments in the elevation of the disk as well as in the angular orientation of the oscillating system.

Fig. 5.5 Kestin disc viscometer oscillating within fixed plates

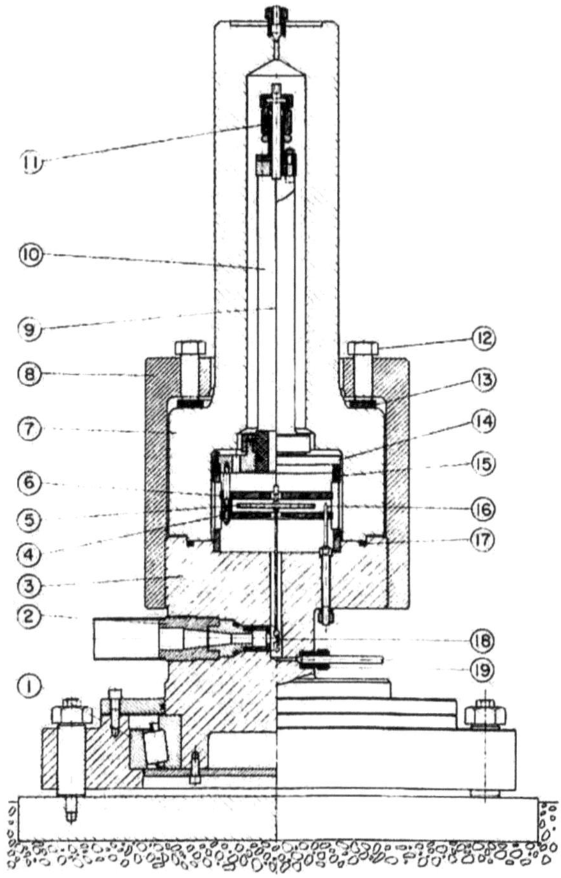

The oscillating disk carries on its lower side a 2 mm diameter Hastelloy stern that is rigidly attached to it. The stem carries at its lower end a small, highly polished mirror 18. The mirror faces towards a 13 Bridgman-type sapphire window assembly 2. The window assembly is screwed into the lower body and is sealed by means of an O-ring.

The suspension cartridge can be lifted from the viscometer simple by loosening a locking screw set in the bottom of the thick-walled cylinder. Prior to filling the suspension cartridge, the disk can be immobilized by inserting small pieces of Teflon between the disk and the lower fixed plate. This feature simplifies the suspension cleaning and alignment operations and ensures protection of the suspension system when cleaning the lower part of the viscometer. The alignment or the suspension is checked by means of a precision theodolite.

5.8 Viscometer Used by Berstad et al.

The viscometer used was essentially the same as used by Torklep and Oye [13] but design changes were made (1) to reduce tolerance, (2) to improve rigidity and (3) to lower the sensitivity to external vibrations. The viscometer was fully computerised. The new design improved the reproducibility by a factor of ten and accuracy by a factor of two.

Improvements were made in respect of

(1) Container and temperature control of the sample water
(2) Torsion strand and pendulum
(3) Temperature measurements.

5.8.1 Sample Container and Temperature Control

The sample container is shown in Fig. 5.6. It consists of an inner Pyrex glass vessel E and an outer plexi-glass container D. A double walled brass tube B is mounted on it. The Pyrex glass contains the sample water in which lower part of the pendulum i.e. the body F oscillates. Temperature controlled water enters the plexi-glass through inlet H and comes out of the apparatus through the outlet A. The temperature of sample water remains constant within 1 mK through this process. Temperature of sample water is measured with a calibrated Platinum resistance thermometer. The room temperature is kept under control within 0.5 °C of the sample temperature. The entire oscillating assembly is stabilised close to the sample water. By traversing the Platinum resistance thermometer vertically as well as horizontally, the temperature was found to be constant within ±1 mK both in space and time. Time of measurement was about 25 min to complete one measurement cycle. Other parts are the pendulum C; Pt resistance thermometer G and K is the tube for loading sample water.

The period of oscillation T and logarithmic decrement are measured with a method of timed intervals. A laser beam is reflected from a mirror on the pendulum to sweep across two fixed photo detectors. An accurate watch connected to the detectors times the interval between detectors intercepts, from which decrement and the period of oscillation can be uniquely determined. Moment of inertia and other required dimensions of the oscillating body are measured separately. Equation (5.15) rather than (5.16) is used for calculation of viscosity.

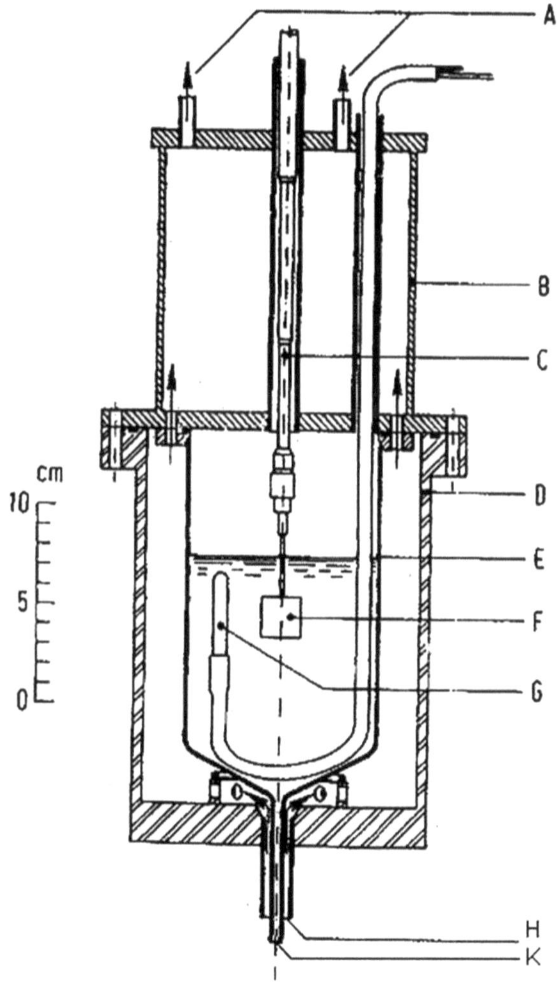

Fig. 5.6 Schematic diagram of sample container

5.9 NBS Torsion Pendulum

5.9.1 Torsion Pendulum

A torsion pendulum essentially consists of a heavy bob suspended with a wire from a rigid fixed point. When the bob is excited in torsion oscillation about its vertical axis, it executes simple harmonic motion whose time period depends upon the torsion couple of the wire, moment of inertia of the bob and upon the viscosity of the surrounding media.

With vacuum outside the bob, its time period is given by

$$T = 2\pi\sqrt{\frac{I}{k}} \tag{5.26}$$

Here I is moment of inertia of the bob and k is torsion constant of the wire. In case, the bob is hollow and is filled with a fluid or the bob is rotating in a fluid, in either case, its time period will change and its maximum amplitude will be decreasing continuously. In this situation, it may be used as a viscometer. White and Kearsley [16] expressed kinematic viscosity v as

$$v = R^2\sqrt{\frac{k}{I_O}}f(\tau, \alpha) \tag{5.27}$$

where

$$\tau = \sqrt{\frac{k}{I_O}} \cdot \frac{T}{2\pi} \text{ and} \tag{5.28}$$

$$\alpha = \frac{I}{I_O} \tag{5.29}$$

where v is the kinematic viscosity, k is the torsion constant of the suspension, R is the internal radius of the hollow sphere, T is the period of the oscillation, I is the moment of inertia of the empty sphere, and I_O is the moment of inertia of the test fluid considered as a rigid sphere. The function $f(\tau, \alpha)$ is a mathematical function of the two dimensionless variables τ and α hence $f(\tau, \alpha)$ is a dimensionless quantity. An accurate determination of viscosity requires measurements of all the physical quantities entering into these equations. The calculation of $f(\tau, \alpha)$ can be done to the necessary accuracy by numerical methods using a modern digital computer. Details of this calculation are given in [17], (where symbol η is used for function f).

5.9.2 Torsion Viscometer

A torsion pendulum meant for the determination of viscosity used by NBS [14] is shown in Fig. 5.7. It consists of a heavy bob of 10 cm diameter suspended from the wire of 500 cm. It is enclosed in a double walled vessel. Pressure inside the chamber may be decreased to 0.2 mm Hg. To observe the movement of the bob in the horizontal plane, an image of a thin straight wire through the lens system Q is imaged on the mirror P attached to a very thin rod from the polar hub of the bob. The time period is measured electronically.

To avoid the errors in measurement of time period of oscillation of the viscometer caused by vibrations from nearby traffic, which could cause a pendulum

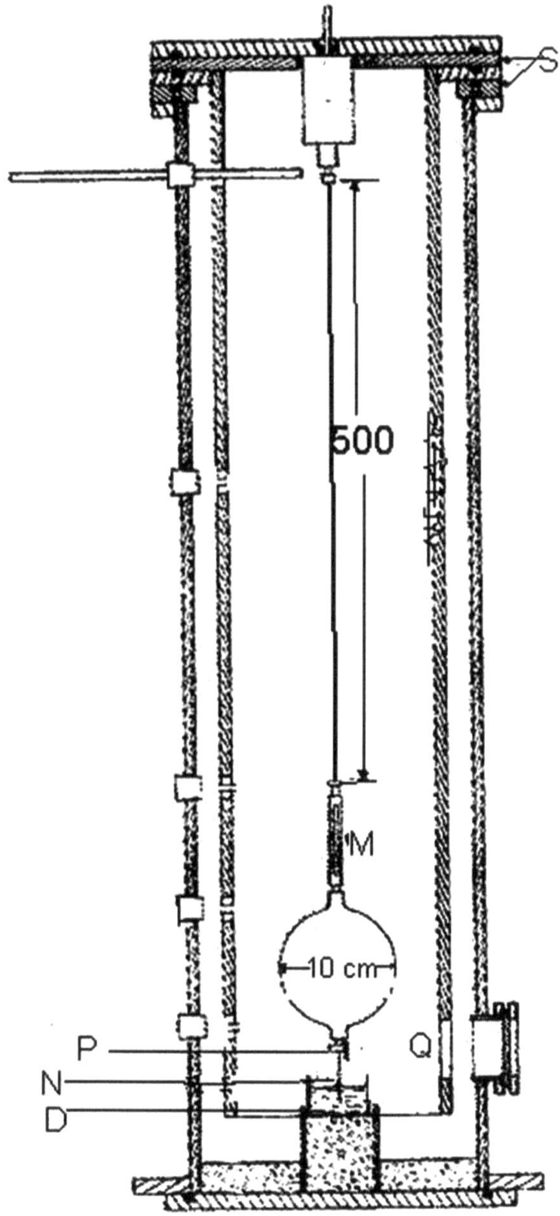

Fig. 5.7 NBS pendulum viscometer

motion of the viscometer in the vertical plane and hence interfere with the measurement of the period. A needle N from the lower polar hub of the sphere (or cylinder), extends into a container of high viscosity mineral oil (viscosity 1000 cSt or $mm^2\ s^{-1}$ at 25 °C). This effectively damped the swinging motion with no measurable effect on the period of the oscillation. This arrangement is used when calibrating the torsion wire.

5.9.2.1 Suspension Wire and Its k Value

The suspension wire was made of an alloy of 92 % platinum and 8 % tungsten with length of 50 cm and diameter of 0.051 cm. The suspension wire was heat treated to about 704.4 °C (1300 °F). The calibration of the viscometer requires an extremely accurate evaluation of the torsion constant k. It is, therefore, necessary to measure the constant at various temperatures and for various tensile loads on the wire. The k value of the wire is determined by measuring time periods of several cylinders in partial vacuum (0.2 mm Hg pressure). White and Kearsley found steal and brass stocks unsuitable for making these cylinders, presumably because these materials were not sufficiently homogeneous. Special Aluminium alloy 6061 and oxygen-free high-conductivity copper were found to be suitable materials. Six cylinders were made with moments of inertia ranging from about 11,000 to 1,200 g cm^2 and weights ranging from 2,100 to 510 g. The measurements of period were made with the cylinders oscillating in torsion while suspended in the position of the sphere shown in Fig. 5.7. The moment of inertia should be calculated from the accurately known dimensions of the cylinder made of homogenous material. The dimensions of the cylinders should be measured within 0.5 μm and mass within 10 mg. If the measurements are made with aforesaid accuracy, the error in the moment of inertia, even of the smallest cylinder, will cause an error of at most ±0.004 % of the torsion constant. A 10 % error in the calculation of the temperature correction may cause an error of ±0.002 % in moment of inertia. The largest error in the calculation of the moment of inertia comes, however, from the calculation of the correction for the drilled holes, brass fittings, wax used to affix the mirror, etc. This correction amounts to 1.3 g cm^2, 10-% error in this calculation, which is a reasonable amount, would cause an error of ±0.010 % in the moment of inertia of the smallest cylinder.

The temperature of the torsion wire was taken to be that indicated by the mercury in glass thermometers through the thermal jacket shown in Fig. 5.7. To ensure that the apparatus was in thermal equilibrium, the temperature was kept constant at the desired value for at least 24 h before measurements were made. The torsion constant of the wire was found to be

$$k = 11471.5 - 0.0208\,\mathrm{g} - 2.6(t - 25) \tag{5.30}$$

The accuracy in the evaluation of k was better than 0.015 %. Here t stands for temperature.

Errors of less than 5 μs in the measured period and errors of less than 0.05 °C in the temperature assigned to the wire caused a negligible error in the torsion constant. Temperature was maintained within ±0.05 °C and time period was measured within an accuracy of ±5 μs.

5.9.2.2 Sphere (Bob)

To make the oscillating sphere, an Aluminium sphere by machining and grounding with a tolerance of 25.4 μm (0.001 in) was made, it was then fitted with nickel polar hubs concentric to the same tolerance. The sphere was then electroplated with a flash plating of copper followed by a plating of nickel to a thickness of about 2.5 mm (0.1 in). The Aluminium sphere was removed by washing with caustic solution flowed through holes in the polar hubs. The exterior of the sphere was then turned down so that the final sphere had a wall thickness of 2.2 mm (0.085 in) and a nominal diameter of 10 cm. A rough calculation showed that these walls were easily stiff enough to prevent a significant change of shape under load.

Radius of the sphere was calculated by filling air free distilled water up to a fixed fiducial mark in the sight glass at the top of the sphere and measuring its mass. The sphere, during this period, was kept immersed in a bath of oil maintained at 25.00 °C. Internal volume of the sphere was calculated by applying all necessary corrections for density, buoyancy and accounting for the volume of water in the sight glass and in the holes in the polar hubs.

The measured internal volume of the sphere at 25 °C was 509.515 cm^3 ± 0.002 %, which is the volume of a sphere of radius 4.9548 cm. The uncertainty of ±0.002 % assigned includes the spread of a number of weighing and repetitions with two oils of known density.

The value for the equatorial radius of the sphere was also found out by filling the sphere with an oil of very high viscosity and measuring the time period. This is approximately equivalent to measuring the moment of inertia of a rigid sphere. Oil N with a viscosity of about 1000 mm^2/s was used for this purpose. The moment of inertia was calculated from (5.26) and the viscosity of oil was measured by a capillary viscometer. This value of viscosity need not be extremely accurate since 1 % error will lead to an error in effective radius of only about 0.005 %. The equatorial radius measured in this way was 4.95525 cm. The difference between this and the radius of a sphere equal in volume to that measured is less than 0.01 % and this difference may be taken as the uncertainty in the radius of the sphere. In fact, the evidence was that the figure of the "sphere" was close to that of a prolate spheroid with the equatorial radius slightly greater than the polar radius. Consequently, in all subsequent calculations of moment of inertia for any fluid was calculated from the mass of the fluid required to fill the fiducial point and the measured equatorial radius.

5.9.2.3 Moment of Inertia of the Empty Sphere

The moment of inertia of the empty sphere was found to be 4074.45 g·cm^2 by timing the period of torsion oscillations at 25 °C in vacuum. Accuracy was ±0.022 %.

This value was based on the average of a number of observations of the period of the empty sphere in vacuum and on the previously determined value of k. An error of ± 0.007 % corresponds to the spread of a number of measurements made at different times.

It is accounted for, in part, by the condensed or absorbed material on the sphere wall which apparently varied each time the sphere was subjected to a vacuum (the measured weight varied by about ± 0.005 %), and, further, by small changes in the mounting of the mirror. There is also a possible error of 0.015 % associated with the value of the spring constant k.

This error will not contribute to the scatter in the observations of period. The actual timing errors involved in measuring the period of the empty sphere were completely negligible, since they amounted to only 2×10^{-4} % of the period.

Equation (5.26) applies to a sphere of fluid contained in a rigid shell of moment of inertia I. Thus, in calculating I, the effective moment of inertia of the small amounts of fluid in hub and sight glass should be added to the moment of inertia of the empty sphere. In practice, these corrections for the nickel sphere were negligible because the small cylinders of fluid are all on the axis of rotation.

5.9.2.4 Measurement of Time period

The period of the torsion viscometer was measured with electronic timer using an optical lever and photomultiplier. Light from a thin (0.1 mm) straight filament of a shielded lamp was made to fall through a window and lens Q to a small galvanometer mirror P shown in Fig. 5.7. The mirror P was fastened to a brass fitting on the lower pole of the sphere. The filament image is reflected through the lens and window to a horizontally adjustable vertical slit 0.13 mm (0.005 in) wide in an opaque film on the outside of a slotted black box enclosing a photomultiplier tube. The output from the photomultiplier tube was connected to the timer and to an oscilloscope. The photomultiplier and electronic timer arrangement was checked against the NBS standard frequency and found to measure time intervals to within 5 μs. Since time periods measured were of several seconds duration, the error in time measurement was completely negligible. The principal difficulties of the measurement of period were found in the mechanical arrangements. Actually, half periods were measured by timing the intervals between zero crossings of the light beam.

At the beginning of the oscillations, there is transient effect, referred to as a “zero·point drift” in the analysis of the apparatus [16], which dies out in a few oscillations. This effect is illustrated in Fig. 5.8 which shows some data for the Aluminium alloy sphere filled with water. The effect is especially large with low viscosity fluids and spheres of low moment of inertia. The time interval between zero crossings is plotted against the number of the zero crossing at the beginning of the interval. The transient effect causes even half periods to be larger than the odd ones. This effect can be used to adjust the slit, first by rough mechanical adjustment and finally by setting the trigger voltage of the timer. When the even

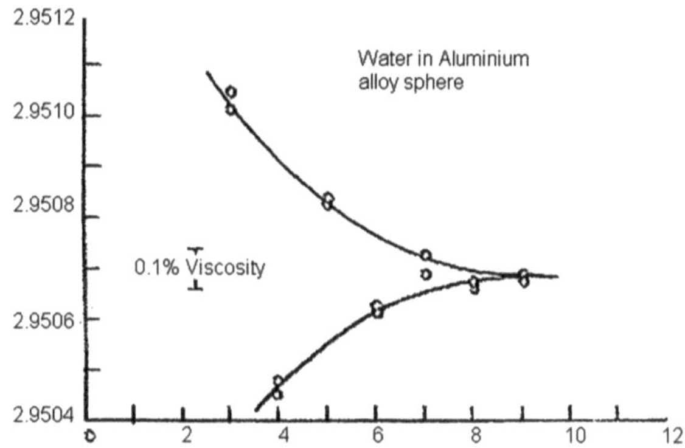

Fig. 5.8 Change in time period (y axis) with number of oscillations passed before start

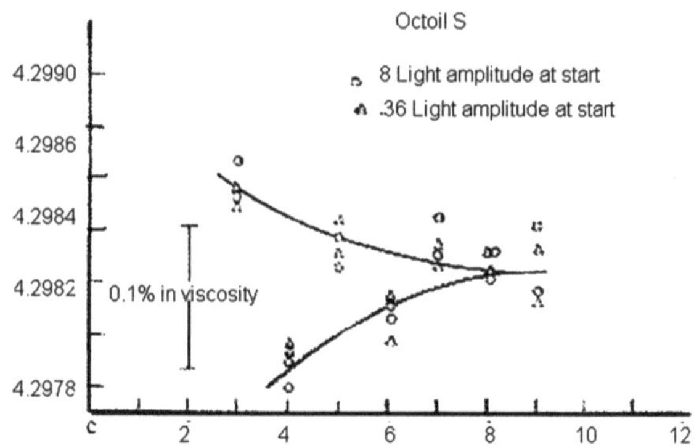

Fig. 5.9 Change in time period (y axis) with number of oscillations passed before start

and odd half periods approach the same limiting value, as they do in Fig. 5.8, the slit is set to give a correct zero position.

It is seen from the Fig. 5.8 that time period may be measured after leaving 10 oscillations. Figure 5.9 shows a similar plot for the nickel sphere filled with Octoil S, but for two different starting amplitudes. No effects of starting amplitude are evident although the starting effect of the previous paragraph is present. This fact is important in establishing that secondary flows due to inertial effects in the fluid are not causing a measurable effect on the period.

But for liquids with a viscosity some 50 times that of Octoil S, a pronounced amplitude effect has been observed. White and Kearsey [16] obtained the data with starting light amplitudes of 12–36° the same temperature and pressure within a 40 min period of time, and these data show an increase of time period with pass number until the amplitude has decreased to about 8°. The data for starting amplitude of the light of 8° (obtained at a different ambient pressure and slightly different temperature) showed no change of period with pass number.

5.9.2.5 Calculation of Kinematic Viscosity

We have seen from the above discussions, for absolute determination of viscosity of a liquid one has to measure the following parameter of the torsion viscometers, which are constant in nature and may be used for number of sample and need to be measured at the time of calibration:

1. b- the effective equatorial diameter of the oscillation sphere
2. k-the torsion constant of the suspension wire
3. I-the moment of inertia of oscillating sphere when empty
4. V-volume of the oscillating sphere

The following parameters are required to be measured for each sample.

5. I_o- the moment of inertia of the liquid filled in the oscillating sphere
6. ρ-density of the liquid under-test
7. T-Time period of the oscillating sphere filled with under-test liquid

I- the moment of inertia is calculated by the time period measurement.

Measured data is as follows

b = equatorial radius = 4.95525 cm, $\delta b/b = \pm 1.0 \times 10^{-4}$

k = torsion constant of wire = 11455.54 dyn·cm and $\delta k/k = \pm 1.5 \times 10^{-4}$

I = measured moment of inertia of empty sphere = 4074.445 g cm^2 and $\delta I/I = \delta k/k$

V = volume of sphere excluding sight glass and hubs = 509.515 cm^3 and $\delta V/V = \pm 2 \times 10^{-5}$

ρ = density of test fluid = 0.91014 g/cm^3 (Air-free Octoil S at 0.24 mm Hg, 24.97 °C)

and $\delta\rho/\rho = \pm 2 \times 10^{-5}$.
Observed period, $T = 4.297637$ s, having negligible measurement uncertainty
The relations used are:

$$I_o = \frac{2}{5} V\rho b^2 = \frac{2}{5} \times 509.515 \times (4.95525)^2 \times 0.91014 = 4554.660 \quad (5.31)$$

$$\delta 1_o/1_o = \delta\rho/\rho + \delta V/V + 2\delta b/b$$

$$\frac{\delta I_o}{I_o} = 2 \times 10^{-5} + 2 \times 10^{-5} + 2 \times 1.0 \times 10^{-4} = 2.4 \times 10^{-4} \quad (5.32)$$

$$\alpha = \mathrm{I}/\mathrm{I}_o = 0.894566 \text{ and } \tau = \sqrt{\frac{k}{Io}} \cdot \frac{T}{2\pi} = 1.084750 \quad (5.33)$$

$\nu = R^2 \sqrt{\frac{k}{Io}} f(\tau, \alpha)$ [see (5.27)].

In the table giving the calculated values of $f(\tau, \alpha)$ [16], there is a line with $\alpha = 0.89450$, $\tau = 1.084737$, giving $f(\tau, \alpha) = 0.0050100$, and in this region of the table $\Delta f/f = 0.07042$, $\Delta\alpha/\alpha = -0.03380$. Thus, a linear interpolation gives the $f(\tau, \alpha)$ corresponding to the measured values τ and α as 0.0050086.

Substituting the values of b, k, I and $f(\tau, \alpha)$ in (5.27), the kinematic viscosity v is calculated as

$$v = f(\tau, \alpha) \cdot b^2 \sqrt{\frac{k}{I_o}} = 19.504\,\mathrm{mm}^2/\mathrm{s}\ (\mathrm{cSt}) \tag{5.34}$$

Partial differential of (5.27) with respect of different parameters, we get

$$\begin{aligned} \frac{\delta v}{v} = &\left\{1 - \frac{\tau}{f(\tau, \alpha)} \frac{\delta f}{\delta \tau} - \frac{2\alpha}{f(\tau, \alpha)} \frac{\delta f}{\delta \alpha}\right\}\left\{\frac{\delta b}{b} + \frac{\delta \rho}{\rho}\right\} \\ &+ \left\{1 + \frac{\tau}{f(\tau, \alpha)} + 2\frac{\alpha}{f(\tau, \alpha)} \frac{\delta f}{\delta \alpha}\right\}\left\{\frac{\delta k}{k} + \frac{\delta V}{V}\right\} \end{aligned} \tag{5.35}$$

Substituting the value of various parameters from above we get

$$\frac{\delta v}{v} = \pm 6.2 \times 10^{-4}. \tag{5.36}$$

5.9.3 Theory for Calculations of Viscosity

The calculation of the period as a function of the viscosity of the liquid inside the bob is done in two steps: (1) the fluid flow problem is solved and the viscous torque on the shell is calculated; (2) the calculated viscous torque is inserted into the dynamic equation of the shell and the motion of the shell is then found out.

5.9.3.1 Viscous Torque

Taking centre of the oscillating sphere as origin and using the spherical co-ordinates Fig. 5.10, the flow problem is solved in the approximation of small angle of rotation of the shell so that quadratic inertial terms can be ignored. The physical meaning of this approximation is that the effects of fluid flow in planes containing the axis of rotation of the pendulum are ignored. If the pendulum is to serve as an absolute viscometer, this error is estimated later on.

The Navier-Stokes equation in spherical coordinates, ignoring quadratic inertial terms and taking account of symmetry, is given as

$$\frac{\delta w}{\delta t} = v\left\{\frac{1}{r^2}\frac{\delta}{\delta r}\left(r^2\frac{\delta w}{\delta r}\right) + \frac{1}{r^2 \sin\theta} \cdot \frac{\delta}{\delta \theta}\left(\sin\theta \cdot \frac{\delta w}{\delta \theta}\right) - \frac{w}{r^2 \sin^2\theta}\right\} \tag{5.37}$$

where w is the fluid velocity perpendicular to planes containing the axis of rotation (the ϕ component of velocity), and v is the kinematic viscosity. This is a linear

Fig. 5.10 Co-ordinate system

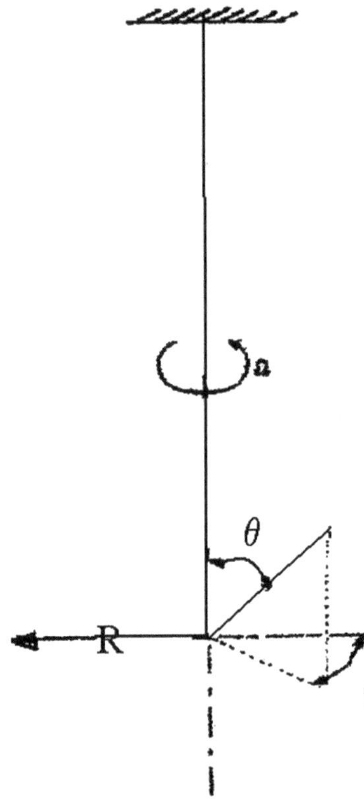

homogeneous equation in w and the time dependence can be separated out by assuming

$$w = w_o \exp(i\Omega t) \tag{5.38}$$

The applicable boundary conditions are:

At r = R, $w_o = i\,\Omega\,\Theta\,R\sin\theta$, where w_o is the velocity of liquid layer in contact of the wall of the sphere at the latitude of θ.

The w_o is continuous and finite within the sphere, Θ is the initial angle through which the sphere is turned from equilibrium.

The solution of this boundary value problem is

$$w_o = \frac{i\Omega\Theta R}{x^2}\cdot\frac{\sin zx - zx\cos zx}{\sin z - z\cos z} \tag{5.39}$$

where

$$z = \frac{i\sqrt{i\Omega}}{\sqrt{v}}R \tag{5.40}$$

$$x = \frac{r}{R} \tag{5.41}$$

The complex parameter z is closely related to boundary layer thickness. Finally the viscous torque L due to this flow pattern acting on the shell is given by

$$L = \left\{\frac{8}{3}\pi R^3 \eta \cdot i\Omega\Theta \exp(i\Omega t)\right\} \cdot \left\{\frac{z^2 \sin z}{\sin z - z\cos z} - 3\right\} \tag{5.42}$$

where η is the coefficient of dynamic viscosity; that is the product of ρ—the density of the liquid and ν its times kinematic viscosity.

5.9.3.2 Dynamic Equation of the Shell

The dynamic equation of the shell is obtained by equating the torques acting on the shell to the rate of change of angular momentum of the shell, viz.:

$$I\Omega^2\Theta \exp(i\Omega t) = k\Theta \exp(i\Omega t) + L \tag{5.43}$$

where I is the moment of inertia of the shell and k is the torque constant of the suspension wire. Substitution of (5.42) into (5.43) and some arithmetic manipulation results in the following equation,

$$-I\Omega^2 + i\Omega\left\{\frac{8}{15}\pi R^3 \rho z^2 \cdot \nu\right\}\left\{\frac{5}{1 - z\cot z} - \frac{15}{z^2}\right\} + k = 0 \tag{5.44}$$

Now writing the first factor of the second term of (5.44) as

$$i\Omega\left\{\frac{8}{15}\pi R^3 \rho z^2 \cdot \nu\right\} = i\Omega\nu\left\{\frac{8}{15}\pi R^5 \rho\right\}\left\{\frac{z^2}{R^2}\right\} = i\Omega\nu\{I_o\}\{-i\frac{\Omega}{\nu}\} = I_o\Omega^2 \tag{5.45}$$

Dividing both sides of (5.44) by $I_o\Omega^2$, we get

$$\phi(z) = \frac{I}{I_o} - \frac{k}{I_o\Omega^2} \tag{5.46}$$

Here

$$\phi(z) = \left\{\frac{5}{1 - z\cot z} - \frac{15}{z^2}\right\} \tag{5.47}$$

Now defining the function $f(\tau, \alpha)$ as follows:

$$f(\tau, \alpha) = \frac{\nu}{R^2} \cdot \frac{1}{\sqrt{\alpha}} \cdot \frac{1}{\omega_o} = \frac{\nu}{R^2} \cdot \frac{1}{\sqrt{\alpha}} \cdot \sqrt{\alpha} \cdot \frac{T}{2\pi} \cdot \frac{1}{\tau} = \frac{\nu}{R^2} \cdot \sqrt{\frac{I_o}{k}}$$

Giving

$$f^2(\tau, \alpha) = \frac{\nu^2}{R^4} \cdot \frac{I_o}{k} \tag{5.48}$$

Further from (5.40),

$$z^2 = -\frac{i\Omega R^2}{v} \tag{5.49}$$

Giving

$$z^4 = -\frac{\Omega^2 R^4}{v^2} \tag{5.50}$$

Multiplying (5.48) and (5.50) we get

$$z^4 f^2(\tau,\alpha) = -I_o\Omega^2/k \tag{5.51}$$

Hence (5.46) becomes

$$\phi(z) = \alpha + \frac{1}{f^2(\tau,\alpha)\cdot z^4} \tag{5.52}$$

Equation (5.52) may be rewritten as

$$z^4\phi(z) = \alpha z^4 + \frac{1}{f^2(\tau,\alpha)} \tag{5.53}$$

In (5.53) α and $\frac{1}{f^2(\tau,\alpha)}$ are real but $z^4\phi(z)$ and z^4 are complex quantities. So writing (5.53) in real and imaginary parts, we get

$$\mathrm{Re}\,\{z^4\phi(z)\} + \mathrm{Im}\,\{z^4\phi(z)\} = \alpha\,\mathrm{Re}(z^4) + \alpha\,\mathrm{Im}(z^4) + \frac{1}{f^2(\tau,\alpha)} \tag{5.54}$$

Now equating real and imaginary parts separately, we get

$$\frac{1}{f^2(\tau,\alpha)} = \mathrm{Re}\{z^4\phi(z)\} - \alpha\,\mathrm{Re}(z^4) \tag{5.55}$$

and

$$\alpha = \frac{\mathrm{Im}\{z^4\phi(z)\}}{\mathrm{Im}(z^{4)}} \tag{5.56}$$

From (5.49)

$$z^2 = -\frac{R^2}{v}\left(\frac{2\pi}{T}i - \frac{2\pi}{T}\delta\right) \tag{5.57}$$

Giving us

$$\mathrm{Im}\left(z^2\right) = -\frac{R^2}{v}\frac{2\pi}{T} \tag{5.58}$$

From (5.27)

$$\frac{R^2}{v} = \frac{1}{f(\tau,\alpha)} \cdot \sqrt{\frac{I_o}{k}} \tag{5.59}$$

Giving us

$$\mathrm{Im}\left(z^2\right) = -\frac{1}{f(\tau,\alpha)} \cdot \sqrt{\frac{I_o}{k}} \cdot \frac{2\pi}{T} \tag{5.60}$$

But τ from (5.28)

$$\tau = \frac{T}{2\pi} \cdot \sqrt{\frac{k}{I_o}} \tag{5.61}$$

Combining (5.60) and (5.61), we get

$$\mathrm{Im}(z^2) = -\frac{1}{f(\tau,\alpha)} \cdot \frac{1}{\tau} \tag{5.62}$$

Giving us

$$\frac{1}{\tau} = -f(\tau,\alpha) \cdot \mathrm{Im}(z^2) \tag{5.63}$$

Again from (5.57)

$$\mathrm{Re}\left(z^2\right) = \frac{R^2}{v} \cdot \frac{2\pi}{T}\delta \tag{5.64}$$

Dividing (5.64) with (5.58) we get

$$-\frac{\mathrm{Re}(z^2)}{\mathrm{Im}(z^2)} = \delta \tag{5.65}$$

Here "Re" and "Im" stand for "the real part of" and "the imaginary part of," respectively.

5.9.3.3 Computations of Tables for $f(\tau, \alpha)$

For a given set of parameters, α and $f(\tau,\alpha)$, there is one primitive complex root of (5.50) and an infinity number of real roots. The former corresponds to a damped oscillation and the latter to purely damped components of the motion. The complex root determines the period of the oscillation, and the real roots are of interest only in that they may affect the experimental measurement of period by causing a "zero drift" error.

The computations are carried out by inserting various complex values of z in (5.53) and the associated parameters are computed. After sufficient data is developed in this way, it was possible to make a good guess at the roots of (5.53). The procedure was then reversed; values of α,$f(\tau, \alpha)$ and an estimated root z are fed to the computer. The root z is then corrected by Newton's method and then τ, *D*, etc. are calculated. A typical line of output from the computer will contain the columns labelled as: α, $f(\tau, \alpha)$, Re(z), Im(z), τ and, $D = \delta T/\pi$.

References

1. Kestin J, Newell GF (1957) Theory of oscillating type viscometers I: the oscillating cup. Z Angew Math Phys 8:433–449
2. Beckwith DA, Newell GF (1957) Theory of oscillating type viscometers, the oscillating cup part II. Z Angew Math Phys 8:450–465
3. Azeitia AG, Newell GF (1958) Theory of oscillating type viscometers III, a thin disc. Z Angew Math Phys 9a:97–118
4. Azeitia AG, Newell GF (1959) Theory of oscillating type viscometers IV, a thick disc. Z Angew Math Phys 10:15–34
5. Newell GF (1959) Theory of oscillating type viscometers V, disc between fixed plates II. Z Angew Math Phys 10:450–465
6. Newell GF (1959) Theory of oscillating type viscometers: the oscillating cup part II. Z Angew Math Phys 10:160–174
7. Roscoe R, Bainbridge W (1958) Viscosity determination by the oscillating vessel method II: the viscosity of water at 20 °C. Proc Phys Soc 72:585–595
8. Kestin J, Khalifa HEA (1976) Measurement of logarithmic decrement through measurement of time. Appl Sci Res 32:483–496
9. Berstad DA, Knapstad B, Lamvik M, Skjolsvik PA, Torklep K, Oye HA (1988) Accurate determination of the viscosity of in temperature range 19.5 to 25.5 °C. Physica A 151:246–280
10. Verschaffelt JE (1915) Viscosity of water using sphere as oscillating body. Commun Phys Lab Univ Leiden 148b:17
11. Andarde ENDAC, Chiong YS (1936) On the determination of viscosity by the oscillation of a vessel enclosing a fluid Part 1. Proc Phys Soc 48:247
12. Andarde ENDAC, Chiong YS (1952) Proc Roy Soc A 211:12
13. Torklep K, Oye HA (1979) An absolute oscillating cylinder or cup viscometer for high temperatures. J Phys E: Sci Instrum 12:875–885
14. Kestin J, Paul R, Shankland IR, Khalifia HEA (1980) High temperature, high pressure, oscillating dis viscometer for concentrated ionic solutions. Ber Bunsenges Phys Chemie 84:1255–1260
15. Kestin J, Shankland JR (1981) The free disk as an absolute viscometer and the viscosity of water in range of 25–150 °C. J Non Equilib Thermodyn 6:241–256
16. White Hobart S, Kearsley Elliot A (1971) An absolute determination of viscosity using a torsional pendulum. J Res NBS 75A:541–551
17. Kearsley EA (1959) An analysis of an absolute torsional pendulum viscometer. Trans Soc Rheol 111:69–80

Chapter 6
New Trends in Viscometers

Abstract To fulfil the need of research minded audience, modern viscometers using various physical phenomena are described. The viscometers discussed are based on tuning fork, ultrasonic, plate waves, Love waves, cantilevers and use of optical fibres. A tuning fork generates sound using the phenomenon of resonance at a constant frequency. Similarly a tuning fork-type viscometer resonate its sensor plates at a natural frequency and measures viscosity from the driving force (electromagnetic force) required to maintain constant amplitude. Measurement of acoustic and shear impedances of ultrasonic longitudinal waves are functions of viscosity of the liquid. A viscometer based on aforesaid principle has been described. Martin et al. showed that the attenuation of the plate waves propagating on thin silicon-nitride membranes in contact with viscous liquids depends upon the viscosity of the liquid while the decrease in frequency of maximum transmission depends upon the density of the liquid. Thus an ultrasonic plate wave viscometer along with its basic theory is briefly described. Love waves-plain polarised shear waves using micro devices have been profitably used for viscosity measurement. The Change in frequency of a PZT crystal is a function of the product of viscosity and density of the liquid around it. Piezoelectric resonator has a good potential to measure viscosity and density of the liquids. Such a device is also given. Similarly the micro-cantilevers used for measurement of very small forces can be used to determine density and viscosity of liquids. The change in amplitude and velocity of a vibrating optical fibres partially immersed in a liquid depends upon the viscosity of the liquid. The change in intensity of the diffraction pattern is proportional to the small changes in amplitude of the vibrating optical fibre. This phenomenon has been used to determine viscosity of liquids available in very small amount.

6.1 Tuning-Fork Viscometers

A tuning fork generates sound using the phenomenon of resonance at a constant frequency. Similarly a tuning fork-type viscometer resonate its sensor plates at a natural frequency and determines viscosity from the driving force (electromagnetic

S. V. Gupta, *Viscometry for Liquids*, Springer Series in Materials Science 194,
DOI: 10.1007/978-3-319-04858-1_6,

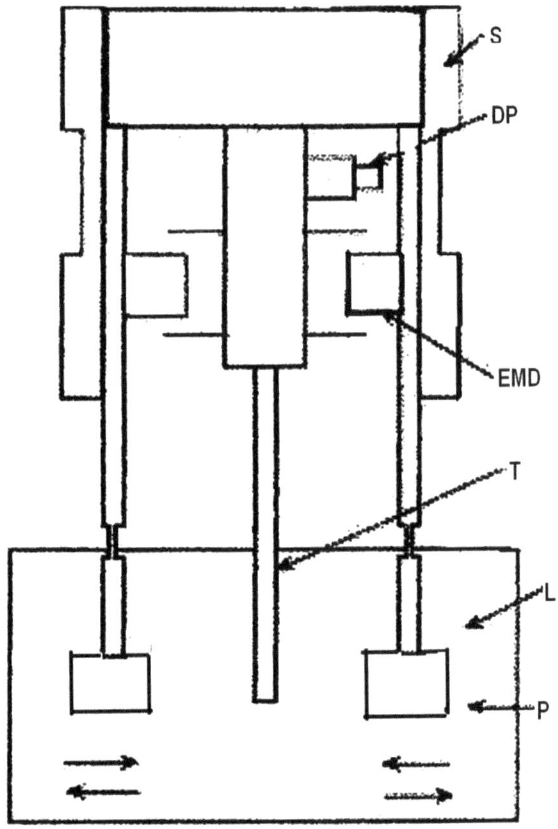

Fig. 6.1 Tuning fork viscometer

force) required to maintain constant amplitude. Such a viscometer gains its high sensitivity from its tuning fork structure. A tuning fork viscometer is shown in Fig. 6.1.

Here S is a spring plate; DP is displacement sensor; EMD is a electromagnetic device; T is temperature sensor; P are vibrating plates and L is liquid under-test in which the plates are vibrating.

6.2 Ultrasonic Viscometer

ANL's ultrasonic viscometer [1] is a non-intrusive on-line device that measures both fluid density and viscosity. The principle of the viscometer is based on acoustic- and shear-impedance measurements, a technique first applied by Moore and McSkimin [2] to measure dynamic shear properties of solvents and polystyrene solutions. Incident ultrasonic shear (1–10 MHz) and longitudinal waves (1 MHz) are launched to two wedge surfaces that are in contact with the fluid, and their reflections are measured. The reflection coefficients, along with sound speed in the fluid, are used to calculate the fluid density and viscosity. Oblique incidence is commonly used because of better sensitivity, but mode-converted waves often occur in wedges that are not of perfect crystal structure and that lack well-polished surfaces. For practical applications, the normal incidence arrangement is used.

6.2.1 Longitudinal Waves and Acoustic Impedance of Fluid

Acoustic impedance of fluid Z_f, is the product of fluid density ρ and phase velocity, V of sound in fluid; it can be determined by measuring the reflection coefficient, R, at the boundary of the fluid and transducer wedge. For normal incidence, R is given as

$$R = \frac{Z_f - Z_w}{Z_f + Z_w} \tag{6.1}$$

where Z_w is the acoustic impedance of the wedge in which longitudinal waves propagate from transducer to fluid. If the phase velocity V in fluid can be determined accurately by other measurement (such as time-of-flight of longitudinal waves travelling in the fluid), the fluid density can be derived from

$$\rho = \frac{Z_w(1 - |R|)}{V(1 + |R|)} \tag{6.2}$$

Here the absolute value of the reflection coefficient is used because in principle R is a complex number. However, in practice, if we assume that wave attenuation in the wedge and fluid can be neglected, only the real parts of R and Z_w are used in the density calculation

6.2.2 Shear Waves and Shear Impedance of Fluid

Use of the ultrasonic shear reflectance method to obtain the shear mechanical properties of fluids has been the subject of many studies for Newtonian [3] and non-Newtonian [4] fluids.

Consider that gated shear-horizontal (SH) plane waves propagate in a wedge at a normal incidence to the polished surface that is in contact with the fluid and are reflected back. The shear reflection coefficient can be expressed as given in (6.1) with shear impedances replacing acoustic impedances. The shear impedances of the wedge, Z_w, and fluid, Z_f, are given as

$$Z_w = \sqrt{\rho_m C_{44}} \tag{6.3}$$

And

$$Z_f = \sqrt{j\omega\rho\eta} \tag{6.4}$$

Here ρ_m is the density of the wedge material; C_{44} is the stiffness constant of the wedge, ω the radial frequency of the shear wave, and η the fluid viscosity. Using (6.4), we have assumed that the fluid behaves as a Newtonian fluid; more complex expressions are expected for non-Newtonian fluids [5]. The shear impedance of fluid is a complex value consisting of amplitude and phase. The phase change is very small for a single reflection, so we consider only the amplitude variation. The shear reflection coefficient R_s is a measurable quantity and therefore can be used to calculate the product of density and viscosity by:

$$\sqrt{\rho\eta} = \sqrt{\frac{\rho_w C_{44}}{2\omega}} \cdot \frac{1 - 2K^2}{K} \tag{6.5}$$

where

$$K = \frac{1 - |R_s|^2}{1 + |R_s|^2} \tag{6.6}$$

Equation (6.5) predicts the measurement sensitivity and range of the shear reflectance method. Figures 6.2 and 6.3 show the dependence of reflection coefficient on the square root of the product of density and viscosity of the liquid for various operating shear frequencies (Fig. 6.2) and wedge materials (Fig. 6.3), respectively. In principle, lower-shear impedance materials and higher operating shear frequencies provide better sensitivity but smaller measurement range.

Ultrasonic viscometers provide instantaneous and continuous measurement of viscosity by means of ultra high frequency sound waves. The first ultrasonic viscometer was made by Rich-Roth Laboratories Hartford Connecticut USA.

The arrangement of six transducers mounted on Rexolite wedge shown in Fig. 6.4 is reported by Viswanath et al. [6]. Out of six, five are longitudinal transducers namely LT1, LT2, LT3, LT4 and LT5 and transducer ST is a shear wave transducer. The transducers have a centre frequency of 2.25 MHz. When ultrasound from transducer F strikes the wedge-liquid interface, part of it is reflected back towards transducer F and the rest is transmitted into the liquid. Similarly when ultrasound from transducers D (or B) operating in pitch catch mode strikes the interface, some of it is reflected towards E (or C), some mode converts to a shear wave in the wedge and part is transmitted into the liquid. The reflection coefficients of the longitudinal waves give the density of the liquid while that of shear waves gives the product of density and viscosity. The coefficients are measured by comparing the voltage on the receiving transducer when the base is immersed in the liquid under test with that when it is immersed in reference liquid, usually water.

Fig. 6.2 Reflection coefficients versus $\sqrt{\eta\rho}$ for different frequencies

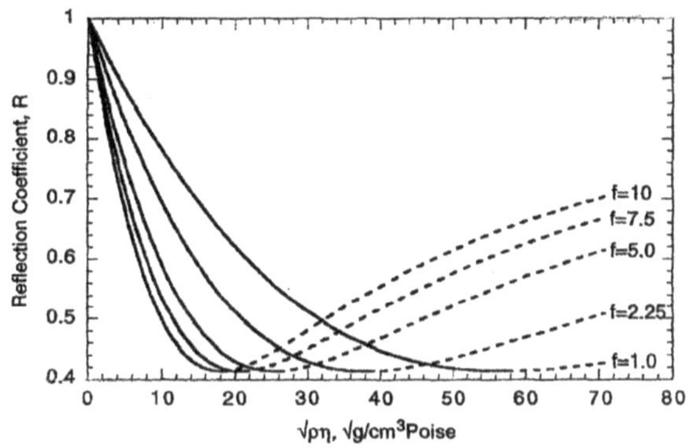

Fig. 6.3 Reflection coefficient versus $\sqrt{\eta\rho}$ for different materials

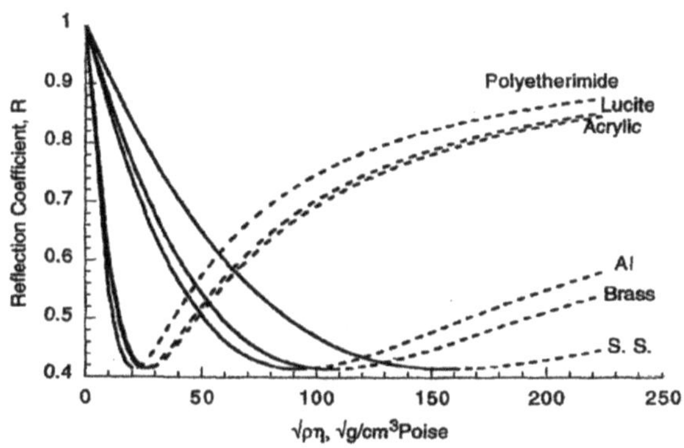

Fig. 6.4 Ultrasonic viscometer (Principle)

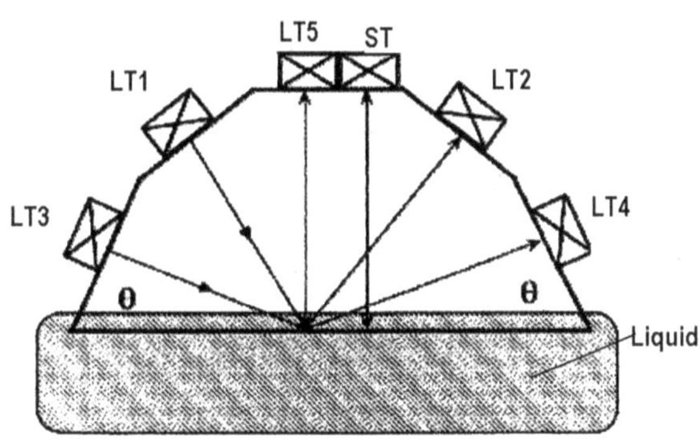

6.3 Ultrasonic Plate Waves Viscometer

In recent years many types of ultrasonic devices have been used for measuring physical properties of liquids [7–10]. Viscosity and density of the liquid are important parameters in using mechanical systems, chemical processes, environmental monitoring and chemical diagnosis. An attractive sensor would be one that measures both of these properties with high sensitivity using small sample volumes. Martin et al. [11] showed that the attenuation of the plate waves propagating on thin silicon- nitride membranes in contact with viscous liquids depends upon the viscosity of the liquid while the decrease in frequency of maximum transmission depends upon the density of the liquid.

6.3.1 Device and Operation

The plate-wave multi-sensor is uniquely suited to liquid-phase measurements. The phase velocity is much lower in thin membranes plate than in bulk solids. For example it is 300–400 m/s in a 3 μm thick membranes plate while its bulk velocity is 3000–8000 m/s in bulk solids. So a membrane plate having thickness smaller than the wavelength of the ultrasonic wave used is chosen. There are two distinct advantages of using a thinner plate, (1) plate-wave devices operate at frequencies that are 10–100 times lower than corresponding bulk-wave or surface-wave devices, and (2) since the phase velocity in a membrane is less than the sound velocity in liquids the device does not radiate energy to the liquid so that it can operate in liquids efficiently.

A typical plate-wave device (Fig. 6.5) consists of a membrane supporting a piezoelectric layer and conducting electrodes fabricated on a silicon wafer. The membrane is made by depositing low stress LPCVD silicon nitride on the wafer and then etching away the silicon beneath using concentrated KOH solution. The roughly rectangular membrane is 3 mm × 8 mm and has angled ends to reduce coherent reflections Aluminium is deposited onto the top of the membrane to form the ground plane. Thickness of Aluminium layer is 0.3 μm. Then zinc oxide is sputtered to form a piezoelectric transducer film, and a second layer Aluminium is deposited over the zinc oxide and patterned to form inter-digitalised transducer electrodes. The transducers contain 50 finger-pairs having 100 μm periodicity, and 2.5 mm apertures. The 5 mm centre-to-centre path length between the transducers sets the wave-liquid interaction length. The thicknesses of the composite membranes are from 28 to 60 μm. The fabricated plate-wave device is mounted with its zinc oxide film down so that the etch pit can be used as a well to contain the liquid. The well is 250 μm deep (set by the thickness of the silicon die) and the volume of liquid in the well is 6 μl. Thus, only a single drop of liquid is needed for these experimental measurements. The device operating in contact with a drop of liquid are made using the membrane as a passive acoustic delay line. A pulsed radio frequency signal is applied to the input transducers and its frequency is tuned to give the maximum output signal, which is displayed on an oscilloscope. The pulse is turned off during the detection of the output signal to eliminate electromagnetic feed through interference. The transmission loss through the device, L is given by

$$L = -20\,\log\left(\mathrm{V_{output}}/\mathrm{V_{input}}\right), \tag{6.7}$$

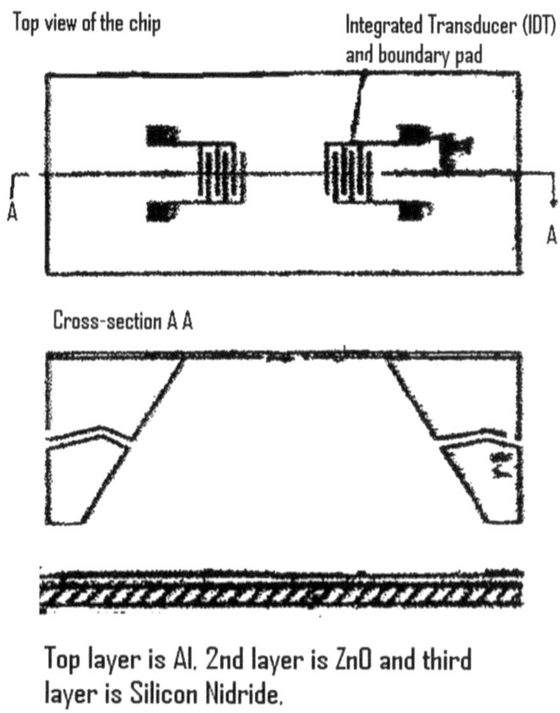

Fig. 6.5 Plate wave device

where, V_{output} and V_{input} are the delay line's output and input voltages respectively and L is in dB.

Since the frequency and the transmission loss signals are obtained simultaneously, the liquid density and viscosity is determination in one measurement.

6.3.2 Basic Theory

6.3.2.1 Viscous Mass Loading

The in-plane component of the A0-mode flexural wave viscously couples with the liquid, causing motion in a viscous boundary layer of thickness $\delta = (2\eta/\rho\omega)^{1/2} = (\eta T/\pi\rho)^{1/2}$, where ω is the angular frequency of the wave and T is its time period.

The liquid, within the viscous boundary layer δ, vibrates with a phase lag and attenuated amplitude with respect to the surface vibrations. The horizontal velocity profile V_x in a liquid near a vibrating plate is given [10] as

$$V_x = V_{x0}\exp(-y/\delta)\cdot\cos(\omega t - y/\delta) \tag{6.8}$$

$$= V_{x0}\exp(-y/\delta)\cdot[\cos(\omega t)\cos(y/\delta) + \sin(\omega t)\sin(y/\delta)] \tag{6.9}$$

The cos(ωt) component of (6.9) is in phase with the surface vibration and contributes (along with the liquid moving perpendicular to the surface) to the mass loading. The out-of-phase sin(ωt) component gives rise to viscous dissipation. To calculate the effective mass, m_l that vibrates in phase with the plate wave, we multiply the in-phase term by the liquid density and integrate throughout the liquid to obtain

$$\begin{aligned} m_l &= \rho \int_0^\infty \exp(-y/\delta)\cos(y/\delta)dy = \rho\delta/2 \\ &= [\rho\eta/2\omega]^{1/2} \end{aligned} \tag{6.10}$$

Using this, the effective mass m_l of the plate-wave velocity expression gives the phase velocity [12]

$$V_p = \left[\frac{B}{M_{total}}\right]^{1/2} = \left[\frac{B}{\left[M_{plate} + \rho\delta_p + (\rho\eta/2\omega)^{1/2}\right]}\right]^{1/2} \tag{6.11}$$

where M_{plate} is the mass density per unit area of membrane, B is the effective stiffness of the membrane, and δ_p, is the skin depth of the pressure wave and is given as

$$\delta_p = \frac{\lambda}{2\pi}\left[1 - \left(V_p/V_r\right)^2\right]^{1/2} \tag{6.12}$$

The corresponding frequency of maximum transmission f is given as

$$f = \frac{V_p}{\lambda} = \frac{1}{\lambda}\left[\frac{B}{\left[M_{plate} + \rho\delta_p + (\rho\eta/2\omega)^{1/2}\right]}\right]^{1/2} \tag{6.13}$$

Here λ is the wavelength set by the transducer periodicity.

This equation is used to calculate the viscous loading curve, which is shown in Fig. 6.6. The dots represent the experimental results. The experimental data fit closely with the theoretical curve, even at high viscosities. Due to viscous mass loading of the membrane, the frequency of maximum transmission is now a function of both density and viscosity. Determination of the Liquid density from the frequency response is possible only if the mass of the viscous boundary layer becomes negligible compared to $\rho\delta_p$, (e.g. for liquids viscosity $\eta < 1$ cP). For liquids with high viscosity, measurement of the transmission loss is needed to account for the viscous mass loading so that density can be accurately determined.

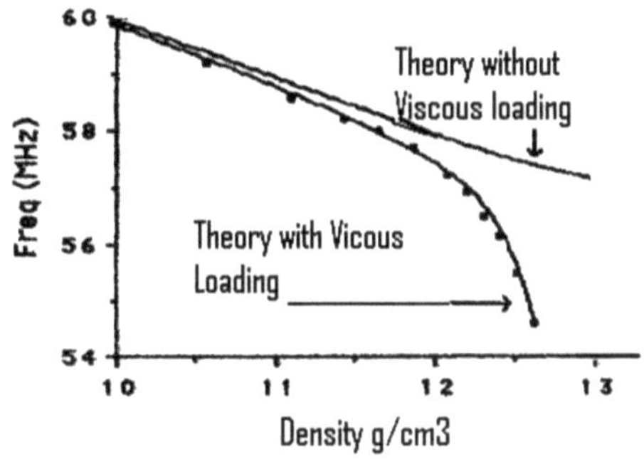

Fig. 6.6 Frequency versus density

6.3.2.2 Viscous Attenuation

The decrease in transmission loss caused by contact with a viscous liquid is found by relating the viscous energy dissipation in the liquid to the acoustic energy carried in the membrane.

The average power dissipation per unit length at the membrane surface is

$$\frac{dP_{loss}}{dx} = \int_0^\infty \eta \left\{\frac{\delta V_x}{\delta y}\right\}^2 wdy = \frac{w}{2}\left\{\frac{\rho\eta\omega}{2}\right\}^2 V_{x0}^2 \tag{6.14}$$

where w is the width of the plate wave on the membrane. The power transmitted along the membrane by the flexural wave is the energy per unit length in the membrane into the group velocity V_g of the wave, giving

$$P = \frac{1}{2} V_m^2 \rho_m w d V_g = V_m^2 \rho_m w d V_p \tag{6.15}$$

where V_m is the mean value of the particle velocity vector in the membrane, ρ_m is the average density of the composite membrane, and d is the membrane thickness. The attenuation coefficient α is then

$$\alpha = \frac{1}{P}\frac{dP_{loss}}{dx} = \frac{(\pi f)^{1/2}}{4\rho_m V_p d} \cdot \left\{\frac{V_{x0}}{V_m}\right\}^2 [\rho_L \eta]^{1/2} \tag{6.16}$$

Since the transmission loss in dB equals the attenuation coefficient integrated over the wave-liquid interaction length, the transmission loss should increase linearly with $(\rho\eta)^{1/2}$. The agreement between theoretical and experimental results can be seen in Fig. 6.7. The graph is in between the frequency on Y-axis and density of the liquid on X axis.

Equation (6.16) shows that the attenuation coefficient α is, to first order, inversely proportional to the membrane thickness d. Thus, thin membranes should be used to increase the measurement sensitivity for low-viscosity liquids ($\eta < 10$ cP) like blood. However, to expand the measurement range to higher viscosities, the

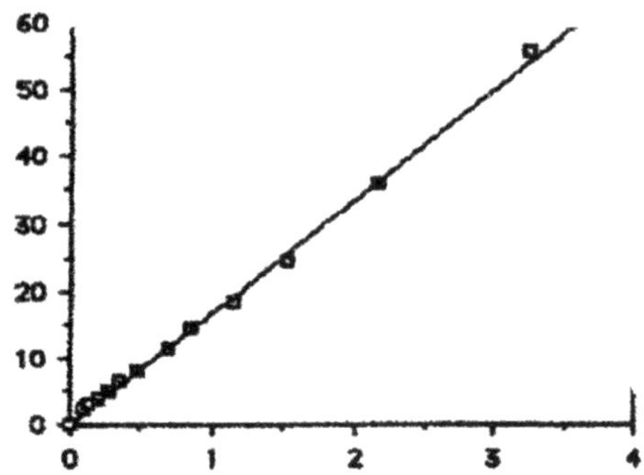

Fig. 6.7 $L/\sqrt{\rho}$ on Y-axis versus $\sqrt{\eta}$ on X-axis L is in dB, ρ in g/cm^3 and η is in mPa·s

devices with thicker membranes should be used. In short, while choosing the thickness of the membrane, some balance is to be kept between sensitivity and maximum measurable viscosity.

The maximum viscosity that can be measured using the plate-wave device depends on the device frequency. At high frequencies, the wave period becomes shorter than the jump-time of molecules due to diffusion in the liquid. If no molecular diffusion occurs during the period of shear strain, energy is stored elastically and no energy is dissipated in the liquid. Thus, viscous attenuation measurements must be done at frequencies that are low enough so that the elastic properties of a liquid may be neglected. Using Maxwell's model of a visco-elastic liquid, the visco-elastic relaxation time τ in a liquid is related to its viscosity by $\eta = \tau$G, where G, is the high-frequency elastic rigidity modulus. Assuming a typical G value of 1×10^9 N/m, one can calculate that a 5 MHz device is good enough to measure for viscosities up to 3×10^4 cP. This range covers most of the oils, lubricants and solvents

In the above calculations, viscous dissipation arising from vertical liquid motion has been neglected. The vertical velocity in the liquid resulting from normal motion on the membrane is

$$V_y = V_{yo} \exp(-y/\delta) \cos(\omega t - 2\pi/\lambda) \tag{6.17}$$

The average power loss per unit length due to shear gradients in V_y is

$$\frac{dP_{loss}}{dx} = \int_0^\infty \eta \left\{ \frac{\delta V_y}{\delta x} \right\}^2 wdy = \frac{w\eta V_{y0}^2}{4\delta}. \tag{6.18}$$

Hence, attenuation due to vertical fled motion is proportional to η Since the measured transmission loss is proportional to $\sqrt{\eta}$, we conclude that the dominant loss mechanism is viscous dissipation with in the viscous boundary layer caused by membrane/Liquid interface.

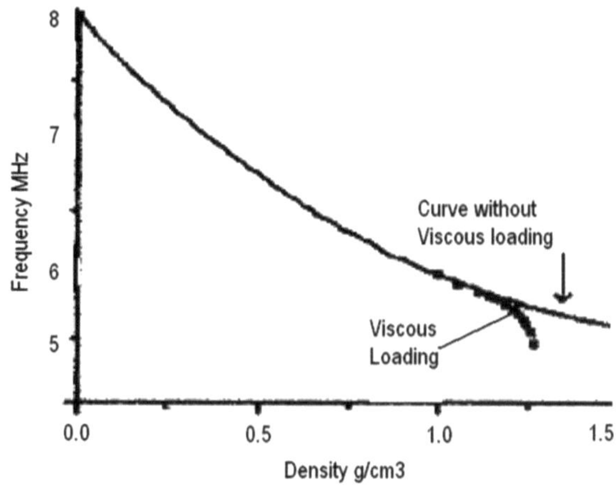

Fig. 6.8 Fall in frequency versus density

6.3.3 Results

Several mixtures of de-ionised water and glycerol having different weight concentrations of glycerol were used. When drops of liquids of different viscosities were placed on a 6 μm thick membrane delay line, the transmission loss *L* in dB divided by square root of density of the liquid is plotted against square root of viscosity. The data are plotted this way because changes m the viscosity of the water-glycerol solutions dominate the response whereas the solution density remains relatively constant. All experiments were performed at 24 °C. From Fig. 6.7, one can see that transmission loss *L* divided by square root of density is proportional to the square root of viscosity. The device responds linearly over a wide viscosity range (from 0.9 mPa·s in water to 1060 mPa·s in 100 % glycerol). The frequency of maximum transducer coupling (Fig. 6.8) decreases by over 2 MHz (25 %) due to loading by the water-glycerol solutions, demonstrating the high sensitivity of the device to density.

6.4 Viscosity by Love Waves

Bernhard and Vellekoop [13] reported an effective way of measuring the viscosity of liquids by using a micro-acoustic Love-wave (horizontally polarized shear waves) device. In elasto-dynamics, Love waves are named after A E H Love and are horizontally polarized shear waves guided by an elastic layer, which is "welded" to an elastic half space on one side, while boarding a vacuum on the other side. Love wave devices provide one of the highest sensitivities among micro-acoustic devices [14–18] and these devices are well suited for liquid-sensing applications. The possibility of viscosity sensing with Love-wave devices has also been discussed. In which the viscosity-induced damping and velocity change of the wave are measured. The chapter [13] gives a suitable electronic sensor set-up, which also allows us to determine whether the test liquid is actually measured in the Newtonian range.

6.4.1 Outline of the Device

The device essentially consists of a thin guiding layer deposited on a substrate, where inter-digital transducers (IDTs) enable electrical excitation and reception of the Love wave. These IDTs are embedded at the interface between the layer and the substrate (see Fig. 6.9). The material for the substrate is piezoelectric ST quartz. The direction of wave propagation is orthogonal to the crystalline X direction. The IDTs, in this orientation, allows the excitation of purely shear polarized waves. As guiding layer, SiO_2 is deposited on the substrate by using plasma-enhanced chemical vapour deposition (PECVD) technique. The acoustic energy associated with the propagating Love wave is concentrated in the guiding layer, which yields a high sensitivity with respect to surface mass loading. The top of the waveguide is the sensitive surface, which is in contact with the liquid being sensed. The sensing surface can be optionally covered with a metallic shielding. The delay line is used as the frequency-determining element in an oscillator circuit; a change in the delay time due to a sensing effect can then be observed by monitoring the oscillation frequency. The sensitivity of Love waves with respect to viscosity has been theoretically discussed by Kovacs et al. [17], which has been briefly discussed in Sect. 6.3.

6.4.2 Advantages of Micro-Acoustic Device

The advantages of using a micro-acoustic device for sensing are:

Real-time electronic readout;
Absence of moving parts;
Small size;
Robustness and
Lesser cost in fabrication.

6.4.2.1 Special Advantage of Love Wave Device

Special advantages for the use of Love wave (shear wave) devices for viscosity sensing are:

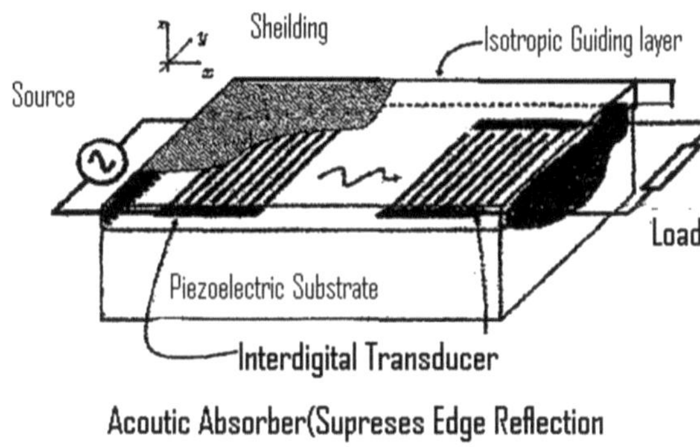

Fig. 6.9 Diagram of love wave device

(a) Due to the pure shear polarization of Love modes, no compression waves are excited in the test liquid and
(b) A high sensitivity with respect to surface mass loading can be achieved [14–16] which, in turn, leads to a very high sensitivity regarding viscosity.

6.4.3 Sensitivity of Love Wave Device

While shear waves show no interaction with ideal liquids, there is an interaction with viscous liquids. Assuming a non-slip boundary condition at the sensitive surface, a thin layer of liquid becomes entrained with the shear movement. The characteristic decay length δ of this entrainment as given by [19] is

$$\delta = \sqrt{\frac{2\eta}{\rho\omega}} \tag{6.19}$$

where η and ρ denote the viscosity and the mass density of the liquid, respectively and ω stands for the angular frequency. The viscous loading affects the Love wave in two ways. First, the entrainment essentially represents a mass loading of the waveguide, which yields a change in the wave number β, which equals $2\pi/\lambda$.

Here λ is the wavelength of the wave.

Secondly, the wave becomes damped due to the viscous losses in the liquid. These effects can be analysed using perturbation theory [19], which yields the following expression for the relative change of the complex propagation constant γ in terms of the unperturbed wave number β) [17]:

$$\frac{\Delta\gamma}{\beta} = \frac{\alpha + i\Delta\beta}{\beta} = (1+i)S\sqrt{\frac{\omega\eta\rho}{2}} \tag{6.20}$$

The real and imaginary parts of $\Delta\gamma$ represent the induced damping α and the change in wave number β respectively. S stands for the device sensitivity [17], which describes the sensitivity of a Love wave with respect to surface mass loading. S directly depends on the scaled over layer thickness h/λ, which is a characteristic quantity of a particular device and depends on its geometry. Since the wavelength λ at the centre frequency of the delay line is prescribed by the IDT period. S is furthermore related to the mass sensitivity S_m, as follows:

$$S_m = S\omega = \frac{1}{\beta} lim \frac{\Delta\beta}{\Delta\rho_s} \quad \text{as } \Delta\rho_s \text{ approaches zero} \tag{6.21}$$

That is, S_m is the relative change of the wave number β due to an infinitesimal surface mass loading $\Delta\rho_s$.

The sensitivity S_m is optimum for a particular value of h/λ. For a fused quartz guiding layer on ST quartz, the optimum value for h/λ is about $= 0.18$ [19].

Equation (6.20) shows that the induced relative damping constant α/β and the relative change in wave number $\frac{\Delta\beta}{\beta}$ are equal to each other and each equals $\sqrt{\frac{\omega\eta\rho}{2}}$, as long as the viscosity η is a pure real number, which is true for Newtonian liquids.

However, liquids tend to show elastic behaviour at high frequencies due to relaxation effect, which is represented by an imaginary part of η. Hence it is necessary to make sure that the frequency and viscosity values are such that the liquid behaves as purely Newtonian. For non-Newtonian liquids relative damping constant and phase change will be unequal. Hence to ensure that viscosity of the liquid is measured for Newtonian range, it is necessary that damping constant and phase change are simultaneously measured and viscosity value is derived when both are equal. Bernhard and Vellekoop have adequately described the method in their chapter [13].

Water glycerol solutions with different glycerol contents have been used to obtain liquids of different viscosities. The experimental results show good agreement with the theoretically derived behaviour. Figure 6.10 shows the measured and calculated damping of a Love wave for liquid loading with glycerol-water solutions having various viscosities. The abscissa $\sqrt{\eta_0\rho\omega}$ is chosen to check for the predicted linear dependence at low viscosities. Note, however, that for the entire measurement $\sqrt{\rho\omega}$ is approximately constant and thus essentially the dependence shown in Fig. 6.10 is on $\sqrt{\eta_0}$ (η_o is the real part of viscosity). Measured points are shown by open circles. The dot-dashed line represents a linear fit through the values measured for low viscosities. The measurement has been performed by covering a well defined area on the propagation path of a Love-wave device with various test liquids.

For further reading, one may consult references [20–23].

Fig. 6.10 Loss in dB per mm versus $\sqrt{\eta_o \rho \omega}$

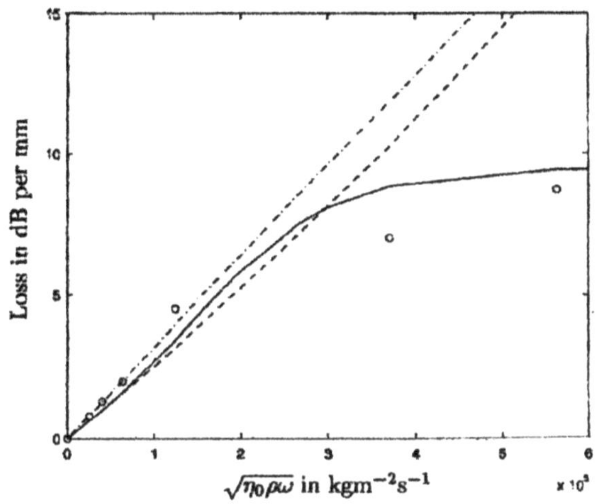

6.5 Piezoelectric Resonator

6.5.1 Change in Frequency Versus Change in Mass

Scientific research is being pursued and papers on acoustic waves, plate wave and cantilevers [24–27] have appeared. Piezoelectric resonators have been used as mass detectors [28]. The following equation is used.

$$\frac{\Delta f}{f} = \frac{\Delta m}{m} \tag{6.22}$$

where Δf is the change in resonant frequency, f is the basic resonant frequency, m is the mass of the resonator and Δm is the surface mass change. The device has been applied to determine gases [29, 30], ions [31], microbes [32], biopolymers [33–35] and electro-chemical mass change [36, 37].

6.5.2 Change in Frequency Versus Viscosity

Studies of quartz crystals in contact with the liquids have also been reported [38–40] by using Kanazawa and Gordon equation [41]. The equation relates the frequency shift with the viscosity, and density of the liquid and shear stress in the quartz and its density as

$$\Delta f = f^{3/2} \left\{ \frac{\eta \rho}{\pi \mu \rho_q} \right\}^{1/2} \tag{6.23}$$

where the viscosity of the liquid is η, μ is the shear modulus of quart, ρ_q is the density of quartz and ρ is the density of the liquid.

6.5.3 Impedance Versus Viscosity

In a study of the electrical resistance of quartz resonators in contact with the liquid at their resonant frequency, an equation has been derived, considering the energy loss on the surface of the quartz crystal [42]

$$R = (2\pi f \rho\eta)^{1/2} \cdot A/k^2 \tag{6.24}$$

Here R is the resistance, which is included as one of the parameters in the electrical equivalent circuit, A is the area of the quartz resonator and k is the electro-mechanical coupling factor of the quartz resonator.

6.5.4 Piezoelectric Resonator in Biochemical Reactions

The Muramatsu et al. [43] used PZT Resonator as a Chemical and Biochemical Sensing Device. The device is used to determine gelation time, phase transition, electrochemical mass change and odorant concentration. Gelation and phase transition are detected as a visco-elastic change, and electrochemical mass change and odorant are detected as a mass change. Their experimental results were in agreement with those obtained by conventional methods.

6.5.5 Quartz Microbalance

The Quartz Crystal Microbalance (QCM) and the electrochemical quartz crystal microbalance (EQCM) are well known methods of viscosity measurement based on the mechanism of resonance. The QCM sensor consists of an AT-cut quartz crystal with two electrodes on opposite sides. Applied electrical field leads to the vibration of the crystal (through the piezoelectric effect). A transverse acoustic wave is generated by this vibration for an AT-cut quartz crystal in thickness-shear mode. This wave propagates across the thickness of the crystal [28].

6.5.6 Piezoelectric Resonator as Density and Viscosity Sensor

Piezoelectric unimorph cantilevers have also been examined as liquid viscosity and density sensors. The cantilever mainly consists of a PZT driving electrode and PZT sensing electrode, permitting direct measurement of the resonance frequency. The shift in the resonance frequency is closely related to the change in the viscosity [27] (glycerol-water solutions were used in the experiment). Multimode dynamic response of composite cantilever beams in various viscous media such as air, water, ultra pure ethanol, and silicon oil has also been investigated for determining viscosity [44]. Agoston et al. [45] evaluated a vibrating cantilever sensor for viscosity measurements of complex organic liquids and compared it with conventional viscometer and micro acoustic sensor. Another sensor application of PZT cantilevers is real-time in-water cell detection [46]. Two cantilevers with different sensitivities were used to detect yeast cells. Previous successes of viscosity measurements with Piezoelectric (namely with QCM and PZT) have noticeable disadvantages. QCM methods utilize quartz crystal, which is far more expensive than PZT and materials being used in other methods. The studies conducted with PZT cantilevers presented an accurate model, but the model only worked in the range of 1–600 mPa·s [27].

Waxman et al. [47] have advocated the use of piezoelectric probes to detect viscosity by way of phase shifts between input and output signals. Some details of the work are given here.

Two pre-stressed PZT composites were used as probes. One probe was used as actuator, which caused vibrations in the fluid. The second probe was used as a sensor to detect these vibrations and turns them into an output voltage. Success was achieved in the range of 0.8–612 mPa·s. In principle the probes are used as an actuator-sensor pair. The probes are mounted facing each other. The distance between the probes was 4 mm and approximately 4 mm of the probes were submerged in the liquid concentrations. A voltage of 1 V_{rms} is applied to the actuator; the second probe acting as sensor receives a vibration wave and turns it into an output voltage.

The sensor probe's output is read by the impedance analyzer and a ratio of output/input signal, or gains, is recorded at each frequency. Along with gain, the phase shift between the actuator and sensor signal is also recorded.

In the case of gain, Fig. 6.11, the frequency of maximum gain increases with the increase in concentrations of glycerine, i.e. with the increase in viscosity. The magnitude of the gain decreases with the increase in viscosity. Exception is the case of gain with 25 % concentration of glycerine.

The frequency of peak phase shift is plotted against glycerine concentration is shown in Fig. 6.12.

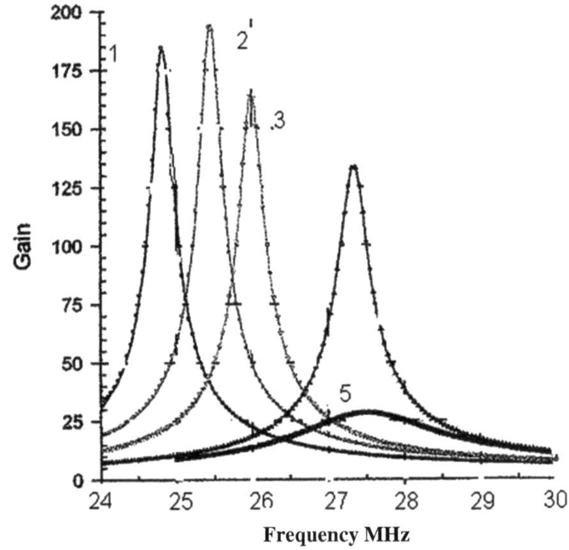

Fig. 6.11 Gain versus frequency for different viscosities

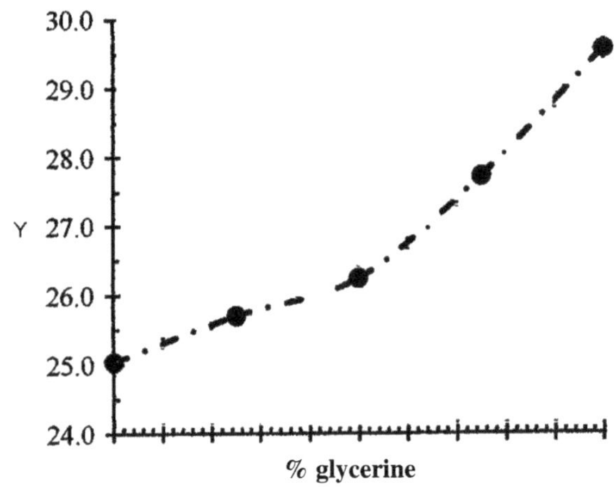

Fig. 6.12 Frequency of peak phase shift versus glycerine concentration

The viscosity and glycerine concentrations are not linearly related. In fact, the viscosity increases exponentially with glycerine percentage [48], viscosity along y axis is plotted against peak frequency of phase shift along x axis is shown in Fig. 6.13.

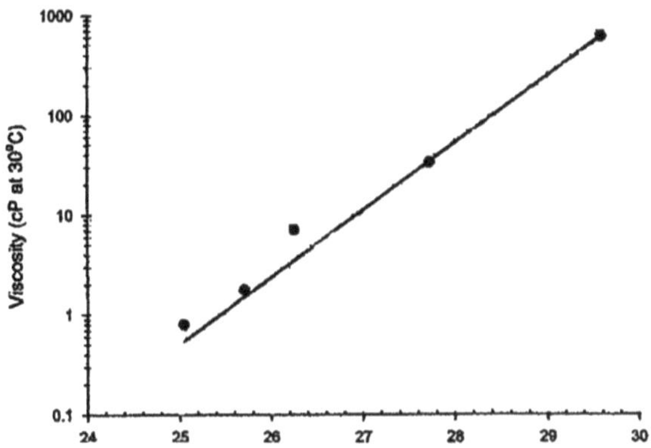

Fig. 6.13 Viscosity versus frequency of peak phase shift

6.6 Micro-Cantilevers for Viscosity Measurement

6.6.1 Introduction

The atomic force microscope (AFM) images the surface of materials in atomic resolution, hence has proved to be a useful tool to investigate mechanical properties and structures in a wide range of materials such as metals, semiconductors, polymers, and synthetic and biological membranes [49, 50]. Similarly it can be profitably used to measure the rheological properties of simple or complex fluids on the nanometre scales [51]. An AFM measures the force, in nanonewton range, in between the microscope tip and the sample surface. The sharp probing tip is attached to a cantilever spring.

Since the frequency spectra of cantilevers strongly depends on the fluid in which they are immersed, the drag force acting on cantilever is directly related to fluid viscosity [52] and its density, hence viscosity can be determined readily from the knowledge of the resonance frequency [53–56] This is a unique feature of such a technique since traditionally information about viscosity is recovered from shear measurements performed on a larger volumes say several millilitres of sample. In mechanical viscometers, viscosity is measured by applying a small amplitude oscillatory shear strain, therefore, can not be used for the study of fragile, rare, and precious materials, including many biological samples that are difficult to obtain in large quantities. There are no such restrictions with AFM. Moreover, while conventional viscometers provide an average measurement of the bulk response, AFM allows for local measurements to be carried out in inhomogeneous systems by probing the material response on micrometer scale. AFM uses a very small volume of sample. A sample of only 10 μl is sufficient. However the determination of the relationship between the frequency responses of a cantilever immersed in a fluid and viscosity poses a formidable challenge [53, 55–58].

For calculation of viscosity, theoretical models do exist. The most simplistic model is based on analogy with the dynamic motion of a sphere through a viscous fluid [59]. This model, although allowing for an analytical solution, fails in many cases since it does not take into account the geometry of the cantilevers [53]. A more rigorous model (in viscid model) [60] overcomes such an insufficiency, but since it does not account for the fluid viscosity it is still inadequate [54]. Finally, a model that describes the dynamic deflection of a cantilever, by rigorously accounting, in the hydrodynamic function, for the geometry of the cantilever and the fluid viscosity has been proposed [54, 55]. Such a model, however, while fully calculating the expected cantilever resonance frequency, thereby allowing for a numerical solution, requires a large amount of numerical computation and a detailed knowledge of the cantilever geometry. A noteworthy contribution was recently developed by introducing a numerical approximation of the hydrodynamic function [61]. This approximation, although considerably simplifying computation, still requires microscopic details of the cantilever geometry that strongly inhibit a wide use of the above model when using commercially available cantilevers. If cantilever geometrical details (length, width, thickness, density, etc.) required by the above models are unknown, a possible alternative approach consists of measuring an empirical AFM-setup calibration curve by using standard liquids such as water-glycerol mixtures at different concentrations or temperatures [62].

6.6.2 Theory

Papi Massimiliano et al. first developed and tested an analytical expression that by using the approximated hydrodynamic function [63], allows for the recovery of samples viscosity upon cantilever calibration. But calibration is a highly time-consuming and subject to several drawbacks for example rupture of cantilevers, they suggested an alternative data analysis approach allowing for the determination of the absolute value of the fluid viscosity by simply accounting for the cantilever resonance frequency measured in the sample under study and in a liquid of known viscosity used as a standard e. g. pure water.

The vacuum resonant frequency ω_0 of a rectangular cantilever is related to the resonance frequency ω in a fluid having density ρ by [53].

$$\omega_0 = \omega\sqrt{1 + \frac{\pi \rho W \gamma_1}{4\rho_0 T}} \tag{6.24}$$

Here ρ_0, W, T are the density, width and thickness of the cantilever and ρ is the density of the fluid. γ_1 is real part of the complex hydrodynamic function (γ), which may be expressed as

$$\gamma = \gamma_1 + i\gamma_2 \tag{6.25}$$

The explicit expression for γ_l and γ_2 are hardly manageable combinations of the Bessel functions and make the use of (6.25) prohibitive. To account for the influence of the fluid motion on the oscillating behaviour of the cantilever an analytical approximation of both the real and the complex parts of the hydrodynamic function has been given [61] as

$$\gamma_1 = a_1 + a_2 \frac{\delta}{W} \tag{6.26}$$

$$\gamma_2 = b_1 \frac{\delta}{W} + b_2 \frac{\delta^2}{W^2} \tag{6.27}$$

where $a_1 = 1.0553$, $a_2 = 3.7997$, $b_1 = 3.8018$, $b_2 = 2.7364$, and δ is the length giving the thickness of the thin viscous layer surrounding the cantilever in which the velocity has dropped by a factor of 1/e and is defined by

$$\delta = \sqrt{\frac{2\eta}{\rho\omega}} \tag{6.28}$$

By substituting the value γ_l and squaring the (6.24), we get

$$\frac{\omega_0^2 - \omega^2}{\omega^2} = \frac{\pi \rho W}{4\rho_0 T} \left\{ a_1 + a_2 \frac{\delta}{W} \right\} \tag{6.29}$$

This gives

$$\delta = \sqrt{\frac{2\eta}{\omega}} = \left[\frac{\omega_0^2 - \omega^2}{\omega^2} - \frac{\pi \rho W a_1}{4\rho_0 T} \right] \div \left(\frac{\pi \rho a_2}{4\rho_0 T} \right)$$

Squaring and expressing explicitly η in terms of other quantities we get

$$\frac{\eta}{\omega} = \alpha \left[\frac{\omega_0^2 - \omega^2}{\omega^2} - \beta \right]^2 \tag{6.30}$$

Here

$$\alpha = 8 \frac{\rho_0^2}{\pi^2 \rho^2 a_2^2} T^2 \text{ And } \beta = \frac{\pi \rho}{4\rho_0} . \frac{W}{T} a_1 \tag{6.31}$$

Writing

$$\frac{\eta}{\alpha\omega} = Y \text{ And } \frac{\omega_0^2 - \omega^2}{\omega^2} = X \tag{6.32}$$

Using (6.32); (6.30) becomes

$$Y = (X - \beta)^2 \tag{6.33}$$

This is the equation of a parabola.

Therefore, by adopting the approximated expression for γ and using a new explicit expression (6.30) one can obtain the sample's viscosity, provided geometrical details of the specific cantilever used are known.

From (6.31) β for any cantilever varies primarily due to change in *W/T*, which normally is fixed by the manufacturer. Moreover it has been shown by Papi et al. [63] that the function does not vary much even if β changes from 1 to 3. While α is proportional to T_2 and hence is more likely to vary from cantilever to cantilever and requires its determination to higher accuracy. Even on the use of scanning electron microscopy (SEM) or a transmission electron microscopy (TEM), dimensional details of the cantilever are only roughly known. So one approach to determine η is to characterize the cantilever elastic behaviour by calibrating it with standards of known viscosity and calculating the values of α and β, which is time consuming and liable to rupture the cantilever.

Another possibility is that we determine resonant frequencies of a cantilever in water and in the liquid under test and utilize (6.32) as follows:

$$Y_s = \frac{\eta_s}{\alpha\omega_s} \tag{6.34}$$

$$Y_w = \frac{\eta_w}{\alpha\omega_w} \tag{6.35}$$

Giving us

$$\eta_s = \frac{Y_s}{Y_w} \cdot \frac{\omega_s}{\omega_w} \cdot \eta_w \tag{6.36}$$

where the parameters with subscript s denote those of sample under test and the parameters with subscript *w* are those of water.

6.6.3 Simultaneous Determination of Density and Viscosity

Neal McLoughlin et al. [64] used water and alcohol to extract parameters characteristic of the resonance behaviour of the cantilever in an AFM system. These parameters were then employed to determine the viscosity and density values of the mixtures. The values obtained were within 5 % of those published in the literature.

In 2008, Papi et al. [65] reported that the vibrating resonance frequency of an un-calibrated atomic force microscope cantilever can be precisely related to the viscosity of the fluid in which it is immersed, independent of any knowledge of the cantilever's geometry and spring constant. Reverse geometry, density, and spring constants of a cantilever immersed in a fluid of known viscosity can be recovered with ease. The proposed methods for monitoring viscosity are of relevance to all bio-rheological and micro-fluidic applications where functionalised cantilevers have to be used, and a simple, yet reliable non-destructive procedure is called for.

6.7 Optical Fibre Viscometer

6.7.1 Introduction

Wei Chih Wang submitted his master's thesis in 1992 [66, 67] suggesting an optical method utilizing the forward light scattering pattern to measure indirectly the viscosity of a fluid. The fluid viscosity was derived from the fluid's damping action on the surface of the immersed vibrating probe (pipette). The resulting viscous fluid motion is referred to as Stokes' (or laminar) flow. Based on a viscous fluid damping model, an expression for the frequency response of the pipette coupled with a fluid was generated. It was shown that the fluid viscosity can be deduced based on a comparison between this expression and the measured maximum vibration amplitude (MVA). This concept was investigated by replacing the pipette with an optical fibre [68]. By vibrating the fibre in the laminar regime, it was shown that different viscous fluids can be identified from the response of the fibre.

Equation of the forced vibration of the optical fibre is given by

$$m\frac{d^2x}{dt^2} + (C_1 + C_2)\frac{dx}{dt} + k(x) = f(t) \tag{6.37}$$

Here *f*(t) is the excitation force from the PZT driver, x is the fibre's displacement, k is the stiffness of the fibre, m is the mass of the fibre and the entrained mass of the surrounding fluid, and C_1 and C_2 are the damping coefficients of the supporting structure and surrounding fluid, respectively. It is convenient to use air as the calibration fluid for each measurement, since it removes the necessity of measuring the exact value of the structural damping of the fibre.

For laminar fluid damping coefficient can be approximated by

$$C_2 = \left\{4\pi f D^2 \eta \rho \pi^2\right\}^{1/2} \tag{6.38}$$

Thus it is worth noting that the fluid damping is proportional to the square root of the product of viscosity η and density ρ of the liquid. Based on this fluid damping coefficient and neglecting the structural damping, the maximum vibration amplitude MVA is given by

$$MVA = \frac{F}{\{4\pi^3 f D^2 \eta \rho\}^{1/2} \omega_o} \tag{6.39}$$

This suggests that for a constant driving-force of amplitude F, the viscosity η can be deduced from the MVA of the fibre provided the mass density of the fluid and the temperature are known.

The sensitivity of the natural frequencies due to changes in fluid viscosity can be made better by partially immersing the fibre in the fluid. Fedorchenko [69] calculated the change in frequency of the partially immersed optical fibre.

6.7.2 Frequency Change of a Partially Immersed Fibre

We know that the wavelengths of a vibrating string of length L fixed at both ends are related by

$$L = n\lambda_n/2 \tag{6.40}$$

Here n is a natural number.

If the speed of wave propagation along the string is c then the angular frequency ω_n or f_n ($\omega_n = 2\pi f_n$) in terms of string length are given as

$$\omega_n = \mathrm{n}\pi c/L \text{ and } f_n = \mathrm{n}c/2L \tag{6.41}$$

In case of a partially immersed fibre, the part immersed in air will be lighter and that immersed in liquid will be heavier due to added mass. So the speed (c) of the wave propagation for the lighter portion and that for heavier portion (c_f) will be different. Similarly the acoustic impedances, which are respectively ρ_c and $\rho_f c_f$, are also different. Here ρ and ρ_f are the densities of the fibre and the fibre with added mass, respectively. In common case this introduces new effect—the running wave reflects not only from the clamped end, but also from the air–liquid interface. As a result, a running wave will interfere with waves reflected from the air–liquid interface and from clamped end. In the present case the ratio of acoustic impedances is close to unity, therefore the intensity of the reflected wave from the air–liquid interface is small and the impact of this wave on the maximum fibre displacement could be neglected.

As to the discontinuity of the speed of propagation, Fedorchenko [69] calculated the constant "effective" speed c_{eff} in such a way that the propagation time of a wave over the fibre length with velocity c_{eff} is equal to the real time of the wave's propagation, that is

$$L/c_{eff} = (L - h)/c + h/c_f. \tag{6.42}$$

It results in

$$c_{eff} = cc_f/[ch^* + (1 - h^*)c_f], \tag{6.43}$$

where $h^* = h/L$.

Using (6.43) and (6.41), the frequencies of the normal modes of the partially immersed fibre become

$$\omega_{eff} = \pi c_{eff}/L$$

$$f_{eff} = c_{eff}/2L. \tag{6.44}$$

Equation (6.44) shows at least qualitatively that the frequency of vibration will decrease with the increase of either density of liquid or the depth of the string immersed in the liquid. To obtain quantitative data about the influence of density, viscosity and liquid depth, one needs to consider forced vibrations for both a fully

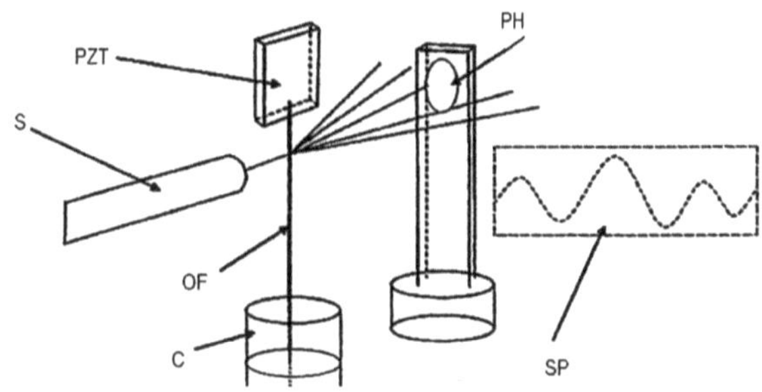

Fig. 6.14 Sketch for the detection of small amplitude

and partially immersed fibre. The maximum vibration amplitude MVA is determined Fedorchenko et al. [70], after a long and laborious mathematics, derived a result that the maximum vibration amplitude MVA as well as the band width are linear function of $(\eta\rho)^{1/2}$ and split over of immersion depth. The band width or MVA for different immersed lengths of the fibre has been measured [70] by the experimental set up shown in Fig. 6.14.

6.7.3 *Experimental Arrangement*

6.7.3.1 Detection of Small Amplitude Vibration

A block diagram of the experimental system is shown in Fig. 6.14. S is a polarized light source from a 10 mW laser, which is incident normally to the optical fibre OF. The fibre is partially submersed in the liquid under-test contained in the container C and is attached at other end to PZT a bimorph Piezoelectric transducer which in turn is connected to a function generator. PH is the photo-detector connected to dynamic signal analyser. The resulting forward scattering pattern from the fibre is indicated as SP. The displacement of the fibre is measured on the intensity variation created by the displacing forward scattering pattern. A linear relationship has been, experimentally observed, between the relatively small fibre displacement (<1 μm) and corresponding intensity variation from the scattering pattern. This provides a simple method of measuring the vibration amplitude of the fibre. In order to deduce the viscosity of the fluid, the fibre's frequency response is determined. This is done by setting the fibre in periodic motion with a piezoelectric transducer. The resulting intensity variation in the forward scattering pattern is measured by a photo detector, and the outputs are collected on a data-acquisition system. Each frequency spectrum of the fibre response is generated by a sufficiently slow sinusoidal input sweep at constant amplitude generated by the piezoelectric transducer. The fluid's viscosity is then estimated by using this spectrum in conjunction with (6.39).

One end of the optical vibrating fibre partially submerged in liquid is fixed to a bimorph piezoelectric transducer while the other end is clamped and secured to the bottom of the stainless steel container by the following ways:

(1) the hole was drilled in the centre of the container's bottom;
(2) a tube of stainless steel was inserted in the hole and fixed with epoxy;
(3) the Fibre has been secured inside this stainless steel tube again with epoxy.

The light from a horizontally placed laser is incident normal to the fibre of diameter *d*, ($d = 125$ μm), density $\rho = 2200$ kg/m^3 and results in a light pattern scattering from the fibre. The bimorph piezoelectric transducer is mounted on a three axis translation stage which allows one to calibrate this device and also to adjust the tension of the optical fibre probe.

In order to deduce viscosity of liquids, a displacement response of the fibre with respect to the driving frequencies is required. It is done by setting the fibre in motion in the vertical plane with a piezoelectric transducer driving at a predetermined angular frequency $\omega = 2\pi f$. This periodic displacement causes an equivalent oscillation of the fibre's forward scattering pattern. The resulting intensity variation caused by the shifting scattering pattern is detected by a photodetector. Obtained data are stored in a computer via a data acquisition system. There is a linear relationship between a relatively small displacement (<1 μm) and its corresponding intensity variation from the scattering pattern.

To generate a frequency spectrum of the vibrating amplitude of the fibre probe, a sufficiently slow sweep of a sinusoidal input at constant amplitude is fed into the piezoelectric driver. Based on change of the resonance curve bandwidth or maximum vibration amplitude (MVA) one can calculate the corresponding damping coefficient of the liquid and, ultimately, deduce the dynamic viscosity of a tested liquid.

It has been shown [70] that variations of the maximum vibration amplitude (MVA) and the bandwidth depend linearly on $(\rho\eta)^{1/2}$ and split over the liquid depth. The two constants of the linear relation are calculated best by method of least squares by measuring either of MVA or bandwidth for liquids of varying viscosity. Experiments were carried out in a wide range of liquid depths and viscosities and confirmed that the density, viscosity and liquid depth in a variable extent impact on the vibration amplitude and bandwidth. The de-ionized water (DI water) and glycerol–water mixtures of different concentrations (GWS) have been used as the testing liquids. The temperature during measurements has been controlled by thermocouple. Experiments have confirmed the method of the viscosity determination.

An analytical solution of the partially submerged fibre vibration with damping has been found and results agree well with experimental data in the wide ranges of the different liquids and depths.

The experimental results [70] as given in Fig. 6.15 show the linear relationship of MVA with $(\eta\rho)^{1/2}$ for different depth h. Similarly linear relation of bandwidth with is seen in Fig. 6.15.

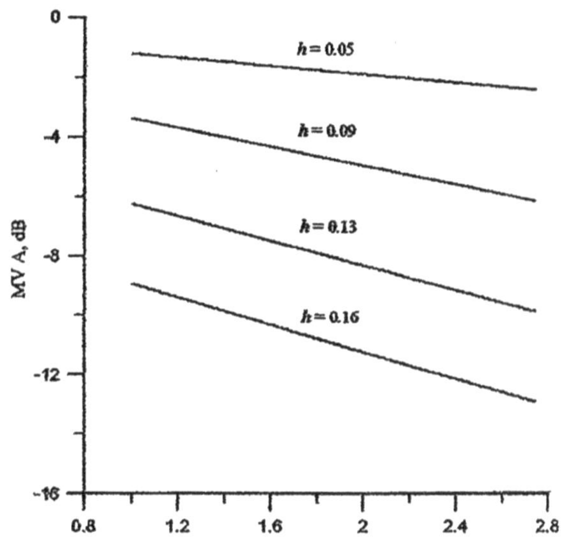

Fig. 6.15 Variation of the MVA with $(\eta\rho)^{1/2}$ for different liquid depths h

Tension F in each case is 1 N
MVA is taken on y axis and $(\eta\rho)^{1/2}$ is taken on x axis
Equations of lines for different values of h are given below:

$y = -0.68x - 0.53$ for $h = 0.05$,
$y = -1.59x - 1.78$ for $h = 0.09$,
$y = -2.08x - 4.17$ for $h = 0.13$,
$y = -2.29x - 6.64$ for $h = 0.16$

Tension F in each case is 1 N.
−3 dB bandwidth is taken on y axis and $(\eta\rho)^{0.5}$ is taken on x axis
Equations of lines for different values of h are given below:

$y = 0.63x + 3$ for $h = 0.05$,
$y = 2.95x + 2.79$ for $h = 0.09$,
$y = 8.12x + 2.74$ for $h = 0.13$,
$y = 16.85x + 2.97$ for $h = 0.16$.

Measurement of viscosity of liquids by tracking the thermal fluctuations of an optically trapped micro-sphere subjected to a small oscillatory flow has been reported by Guzman et al. [71]. The measured power spectral density of the sphere's positions displays a characteristic peak at the driving frequency of the flow, which is simply proportional to the viscosity, when measured in units of the thermal power spectral density at the same frequency. Measurements of viscosity of different water-glycerol mixtures have shown a good agreement with theoretical predictions (Fig. 6.16).

For further reading one may refer papers cited at [72–75].

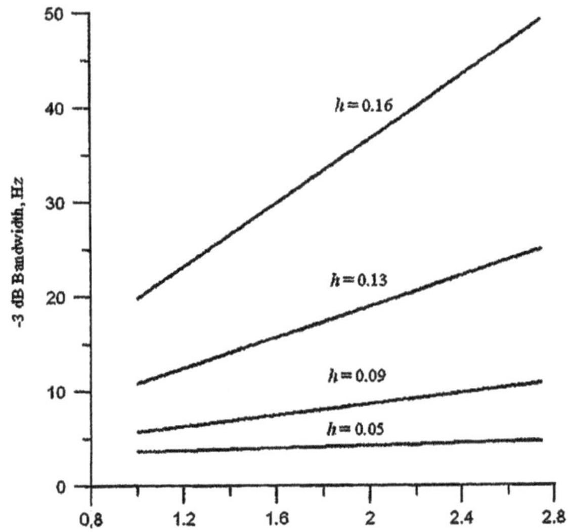

Fig. 6.16 Variation of the −3 dB bandwidth with $(\eta\rho)^{1/2}$ for different liquid depths

6.8 Vibrating Wire Viscometer

In 2005 E Kandil Mohamed submitted a thesis for his Ph.D degree Mohammed [76] described an vibrating wire viscometer to measure the viscosity of fluids over a wide range of temperature and pressure. Viscosity and density of down hole fluids are very important properties as their values can determine the economic viability of a petroleum reservoir. A vibrating wire viscometer has been developed with an electrically insulating tensioning mechanism. Two wires, of diameters 0.05 and 0.150 mm, have been used to measure the viscosity of methylbenzene and two reference fluids with viscosities of 10 and 100 mPa·s at T = 298 K and p = 0.1 MPa, in the temperatures range of 298–373 K and pressures up to 40 MPa, where the viscosity covers the range 0.3–100 mPa·s, with a standard uncertainty <0.6 %. The results differ from literature values by <±1 %. The results demonstrate that increasing the wire diameter increases the upper operating viscosity range of the vibrating wire viscometer.

References

1. Sheen SH, Chien HT, Raptis AC (1997) Ultrasonic methods for measuring liquid viscosity and volume percent of solids A Reported submitted to Argonne National Laboratory, Argonne, Illinois 60439, operated by The University of Chicago for the United States Department of Energy under Contract W-31-109-Eng-38, Report No ANL-97/4
2. Moore RS, McSkimin HJ (1970) Dynamic shear properties of solvents and polystyrene solutions from 20 to 300 MHz. Phys Acoust VI:167–242
3. Cohen-Tenoudji F et al (1987) A shear wave rheology sensor. IEEE Trans Ultrason Ferroelectr Freq Control UEFC-34(2):263–269

4. Harrison G, Barlow AJ (1981) Dynamic viscosity measurement. Meth Exp Phys 19:137–178
5. Sheen SH, Raptis AC (1987) A feasibility study on advanced techniques for on-line monitoring of coal slurry viscosity. Argonne National Laboratory Report ANLEE-87-8
6. Vishwanath DS, Ghosh TK, Prasad DHL, Dutt NVK, Rani KY (2007) Viscosity of liquids—theory, estimation, experiment and data. Springer, Germany. ISBN 978-1-4020-5482-2
7. Wenzeland SW, White RM (1988) A multi-sensor employing an ultrasonic Lamb wave oscillator. IEEE Trans Electron Devices 35:735–743
8. Martin SJ, Ricco AJ (1987) Sensing in liquid using acoustic plate mode devices. In: Proceedings of IEDM-87, pp 290–293
9. Hager HE (1986) Fluid property evaluation by piezoelectric crystals operating in the thickness shear mode. Chem Eng Common 43:25–38
10. Kanazawa KK, Gordon JG II (1985) The oscillation frequency of a quartz resonator in contact with a liquid. Anal Chem Acta 175:99–105
11. Martin BA, Wenzel SW, White RM (1990) Viscosity and density sensing with ultrasonic plate waves. Sens Actuators A21–A23: 704–708
12. White RM, Wenzel SW (1988) Fluid loading of a Lamb wave sensor. Appl Phys Lett 52:1653–1655
13. Jakoby B, Vellekoop MJ (1998) Viscosity sensing using a Love-wave device. Sens Actuators A 68:275–281
14. Kovacs G, Venema A (1992) Theoretical comparison of sensitivities of acoustic shear wave modes for (bio) chemical sensing in liquids. Appl Phys Lett 61:639–641
15. Gizeli E, Stevenson AC, Goddard NJ, Lowe CR (1992) A novel Love plate acoustic sensor utilising polymer over layers. IEEE Trans Ultrason Ferroelec Freq Control 39:657–659
16. Du J, Harding GL, Callings AF, Dencher PR (1997) An experimental study of Love wave acoustic sensors operating in liquids. Sens Actuators A 60:5461
17. Kovacs G, Vellekoop MJ, Haueis R, Lubking GW, Venema A (1994) A Love wave sensor for (bio)chemical sensing in liquids. Sens Actuators A 43:38–43
18. Martin SJ, Ricco AJ, Niemczyk TM, Frye GC (1989) Characterization of SH acoustic plate mode liquid sensors. Sens Actuators 20:253–268
19. Auld BA (1973) Acoustic fields in solids, vol II. John Wiley, New York
20. Kovacs G, Lubking GW, Vellekoop MJ, Venema A (1992) Love waves for (bio)chemical sensing in liquids. In: Proceedings of IEEE ultrasonics symposium, Tucson, AZ, USA, pp 28l–285
21. Jakoby B, Vellekoop MJ (1997) Design of Love wave sensor devices for the operation in liquid environments. In: Proc. IEEE ultrasonics symposium,Toronto, Canada
22. Curtin S, Jakoby B, Berthold A, Varadan VK, Varadan VV, Vellekoop MJ (1998) A micromachined wet cell for a Love-wave liquid sensor. In: Proceedings of smart electronics and MEMS, San Diego, CA, USA
23. Welsch W, Klein CM, von Schickfus M, Hunklinger S (1996) Development of a surface acoustic wave immuno-sensor. Anal Chem 68:2000–2004
24. Martin SJ, Ricco AJ, Hughes RC (1987) Acoustics wave device for sensing in liquids. Transducer 87:478–481
25. Martin BA, Wenzel SW, White RM (1990) Viscosity and density sensing with ultrasonic plate waves. Sens Actuators A: Phys 22:704–708
26. Oden PI, Chen GY, Steele RA, Warmack RJ, Thundat T (1996) Viscous drag measurements utilizing microfabricated cantilevers. Appl Phys Lett 68(26):3814–3816
27. Shih WY, Li X, Gu H, Shih WH, Aksay IA (2001) Simultaneous liquid viscosity and density determination with piezoelectric unimorph cantilevers. J Appl Phys 89(2):1497–1505
28. Lin Z, Yip CM, Joseph IS, Ward MD (1993) Operation of an Ultrasensitive 30-MHz Quartz Crystal Microbalance in Liquids. Anal Chem 65:1546–1551
29. Sauerbrey G (1959) Verwendung von Schwmgquarzn zur Wagung dunner Schlchten und zur Mlkrow & ung. Z Phys I55:206
30. King WH Jr (1964) Piezoelectric sorption detector. Anal Gem 36:1735

31. Hlavay J, Guilbault GG (1977) Application of the piezoelectric crystal detector in analytical chemistry. Anal Chem 49:1890
32. Nomura T, Nagamune T (1983) Internal electrolytic determination of silver in solution with a piezoelectric quartz crystal. Anul Chem Acta 55:231
33. Muramatsu H, KaJiwara K, Tarmya E, Kambe I (1986) Piezoelectric immuno sensor for the detection of Candida albicans microbe. Anal Chem Acta 188:257
34. Shons A, Dorman F, NaJanan J (1972) An imuno-specific microbalance. J Biomed Mater Res 6:565
35. Muramatsu H, bcks JM, Tanuya E, Karube I (1987) Piezoelectric crystal biosensor modified with protein A for determination of immumoglobulins. Anal Chem 59:2760
36. Grabbe ES, Buck RP, Melroy OR (1987) Cyclic voltammetry and quartz microbalance electrogravimetry of IgG and anti-IgG reactions on silver. J Electroanal Chem 223:67
37. Muramatsu H, Tanuya E, Karnbe I (1988) Computation of equivalent circuit parameters of quartz crystals in contact with liquids and study of liquid properties. Anal Chem 60:2142
38. Varmeau PT, Buttry DA (1987) Application of the quartz crystal microbalance to electrochemistry Measurement of ion and solvent populations in thin films of poly (vinyl-ferrocene) as functions of redox state. J Phys Chem 9:1292
39. Nomura T, Okuhara M (1982) Frequency shifts of piezoelectric quartz crystals immersed in organic liquid. Anal Chem Acta I42:281
40. Bruckenstem S, Shay M (1985) Experimental aspects of use of the quartz crystal microbalance in solution. Electrochem Acta 30(10):1295
41. Kanazawa KK, Gordon JG II (1985) The oscillation frequency of a quartz resonator in contact with a liquid. Anal Chem. Acta 175:99
42. Hagar HE (1986) Fluid property evaluation by piezoelectric crystals operating in the thickness shear mode. Chem Eng Commun 43:25
43. Muramatsu H, Suda M, Ataka T (1990) Piezoelectric resonator as a chemical and biochemical sensing device. Sens Actuators A21–A23:362–368
44. Bergaud C, Nicu L (2000) Viscosity measurements based on experimental investigations of composite cantilever beam eigen frequencies in viscous media. Rev Sci Instrum 71(6):2487–2492
45. Agoston A, Keplinger F, Jakoby B (2005) Evaluation of a vibrating micro-machined cantilever sensor for measuring the viscosity of complex organic liquids. Sens Actuators A 123–124:82–86
46. Yi JW, Shih WY, Mutharasan R, Shih WH (2003) In situ cell detection using piezoelectric lead Zirconate Titanate-stainless steel cantilevers. J Appl Phys 93(1):619–626
47. Waxman R, Erturun U, Mossi K (2010) Feasibility of using piezoelectric probes to measure viscosity in newtonian fluids. In: Proceedings ASME 2010 Conference on Smart Materials, Adaptive Structures and Intelligent Systems held on Sept 28–Oct 1, 2010, Philadelphia, Pennsylvania, USA
48. The Dow Chemical Company (2011) Viscosity of aqueous glycerine solutions. 1995–2010
49. Lippert F, Parker D, Jandt K (2004) Caries Res 38(464):2004
50. Perry J, Neville A, Hodgkiess T (2002) A comparison of the corrosion behaviour of WC-CO-Cr & WCCo-HVOF thermally sprayed coatings by in-situ Atomic Force Microscope (AFM). J Therm Spray Technol 11:536
51. C. Bergauda C, Nicu L (2000) Rev. Sci. Instrum. 71, 2487
52. Landau LD, Lifshitz EM (1995) Fluid mechanics, 2nd edn. Pergamon, New York
53. Sader JE (1998) Frequency response of cantilever beams immersed in viscous fluids with applications to the atomic force microscope. J Appl Phys 84:64
54. Sader JE (2000) Experimental validation of theoretical models for the frequency response of atomic force microscope cantilever beams immersed in fluids. J Appl Phys 87:3978
55. Boskovic S, Chon JWM, Mulvaney P, Sader JE (2002) Rheological measurements using microcantilevers. J Rheol 46:891
56. Sader JE, Chon JWM, Mulvaney P (2002) Atomic Force Microscopy in liquids- Biological Applications. Rev Sci Instrum 70:3967

57. Ahmed N, Nino DF, Moy VT (2001) Measurement of solution viscosity by atomic force microscopy. Rev Sci Instrum 72:2731
58. Oden PI, Chen GY, Steele RA, Warmack RJ, Thundat T (1996) Viscous drag measurements utilizing microfabricated cantilevers. Appl Phys Lett 68:3814
59. Chen GY, Warmack RJ, Thundat T, Allison DP, Huang A (1994) Resonance response of scanning force microscopy cantilevers. Rev Sci Instrum 65:2532
60. Chu WH Southwest Research Institute (1963) Elastic Vibration Characteristics of Cantilever Plates in Water. Technical Report No. 2, 963
61. Maali A, Hurth C, Boisgard R, Jai C, Cohen-Bouhacina T, Aima JP (2005) Intrinsically High-Q Dynamic AFM Imaging in Liquid with a Significantly Extended Needle Tip. J Appl Phys 97(074907):2005
62. Buenviaje CK, Ge SR, Rafallovich MH, Overney RM (1998) Atomic force microscopy calibration methods for lateral force, elasticity, and viscosity. Mater Res Soc Symp Proc 552:187
63. Papi M, Arcovito G, De Spirito M (2006) Fluid viscosity determination by means of uncalibrated atomic force microscopy cantilevers. Appl Phys Lett 88(194102–1):194102–194103
64. McLoughlin N, Lee SL, Hähnera G (2006) Simultaneous determination of density and viscosity of liquids based on resonance curves of uncalibrated micro-cantilevers. Appl Phys Lett 89:184106
65. Papi M, Maulucci G, Arcovito G, Paoletti P, Vassalli M, De Spirito M (2008) Detection of microviscosity by using uncalibrated atomic force microscopy. Appl Phys Lett 93:124102-1–124102-3
66. Wang WC (1992) A new optical microsensor for mechanical detection using scattering of a micro-pipette. Master's Thesis, University of washington, Electrical Engineering
67. Wang WC, Afromowitz M, Hannaford B (1994) Technique for mechanical measurement using optical scattering from a micro-pipette. IEEE Trans Biomedical Eng BME-41:298–304
68. Wang WC, Yee SS, Reinhall PG (1995) Optical viscosity sensor using forward light scattering. Sens Actuators B 24–25:753–755
69. Fedorchenko AI, Stachiva I, Ho JN, Wang AB, Wang WC (2008) On the forced vibration of the fibre partially immersed in fluid. Sens Actuators A 147:498–503
70. Fedorchenko AI, Stachiv I, Wang A-B (2009) The optical viscometer based on the vibrating fibre partially submerged in Liquid. Sens Actuators B 142:111–117
71. Guzman C, Flyvbjerg H, Köszali R, Ecoffet C, Forró L, Jeney S (2008) In situ viscometry by optical trapping interferometry. Appl Phys Lett 93:184102–184111
72. Ashkin A (1992) Forces of a single-beam gradient laser trap on a dielectric sphere in the ray optics regime. Biophys J 61:569–582
73. Brau RR, Ferrer JM, Lee H, Castro CE, Tam BK, Tarsa PB, Matsudaira P, Boyce MC, Kamm RD, Lang MJ (2007) Passive and active micro-rheology with optical tweezers. J Opt A: Pure Appl Opt 9:S103–S112
74. Wang WC, Yee S, Reinhall R (1995) Optical viscosity sensor using forward light scattering. Sens Actuators B: Chem 25:735–755
75. Wang WC (1996) A study of Liquid viscosity and flow measurement using fibre-optic transducer. Doct. Dissertation, University of Washington, Electrical Engineering Department
76. Mohammed EK (2005) Thesis for Ph.d degree to the University of Cantebury, Newzealand (Chemical and Process Engineering Department)

Bibliography

1. Perry JM, Neville A, Hodgkiess T (2002) A Comparison of the Corrosion Behavior of WC-Co-Cr and WC-Co HVOF Thermally Sprayed Coatings by In Situ Atomic Force Microscopy (AFM). J Therm Spray Technolo 11(4).

Abstract: The corrosion behaviour of WC-Co-Cr and WC-Co high velocity oxygen fuel VOF) sprayed coatings were examined in static saline conditions. Direct current (DC) polarization tests were conducted and the electrochemical corrosion behaviour was shown, by in situ atomic force microscopy (AFM) and scanning electron microscopy (SEM), to be complex because of the composite ceramic- et al nature of the coating. The addition of chromium to the matrix greatly enhanced the corrosion resistance of the coating.

2. Bergaud C, Nicu L (2002) Viscosity measurements based on experimental investigations of composite cantilever beam eigen frequencies in viscous media. Rev Sci Instrum 71, 2487 (2000);
   Abstract: Experimental investigations have been conducted to study the multimode dynamic response of composite cantilever beams in various viscous media and to determine their viscosity. Theoretical eigenfrequencies are computed using the analytical model proposed by Sader [J. Appl. Phys. 84, 64 (1998)] based on the analysis of the hydrodynamic function of cantilever beams. A good agreement is found between theory and experiment for the first two resonant frequencies of composite beams operated in air and in water. The same experimental approach is used to determine the viscosity of ultra pure ethanol. Thus, it is established that [ISO-8859-1?] Sader's model represents an accurate alternative for the determination of liquid viscosity in small volumes (about 50 μl) which might be of great importance for micro-fluidics applications. Finally, the limits of the method are underlined by monitoring the dynamic response of cantilever beams in silicon oil.
3. Sader JE (1998) Frequency response of cantilever beams immersed in viscous fluids with applications to the atomic force microscopy. J. Appl. Phys. 84, 64 (1998);
   Abstract: The vibrational characteristics of a cantilever beam are well known tostrongly depend on the fluid in which the beam is immersed. In this paper, wepresent a detailed theoretical analysis of the frequency response of a cantilever beam, that is immersed in a viscous fluid and excited by an arbitrary driving force. Due to its practical importance in application to the atomic force microscope (AFM), we consider in detail the special case of a cantilever beam that is excited by a thermal driving force. This will incorporate the presentation of explicit analytical formulae and numerical results, which will be of value to the users and designers of AFM cantilever beams.
4. Chon JWM, Mulvaney P, Sader JE (2000) Experimental validation of theoretical models for the frequency response of atomic force microscope cantilever beams immersed in fluids. J. Appl. Phys. 87, 3978, (11 pages)
   Abstract: Detailed measurements of the frequency responses of a series of rectangular atomic force microscope (AFM) cantilever beams, immersed in a range of fluids, have been performed to test the validity and accuracy of the recent theoretical model of Sader [J. Appl. Phys. 84, 64 (1998)]. This theoretical model gives the frequency response of a cantilever beam that is immersed in a viscous fluid and excited by an arbitrary driving force. Very good agreement between experimental measurements and theoretical calculations is found for all fluids considered. Furthermore, a critical assessment of the well-known inviscid model is presented, which demonstrates that this model is not applicable to AFM cantilever beams in general.
5. Boskovic S, Chon JWM., Mulvaney P (2002) Rheological measurements using micro-cantilevers. J. Rheol. 46, 891, (9 pages)
   Abstract: The use of micro-cantilevers in rheological measurements of gases and liquids is demonstrated. Densities and viscosities of both gases and liquids, which can range over several orders of magnitude, are measured simultaneously using a single micro-cantilever. The micro-cantilever technique probes only minute volumes of fluid (<1 nL), and enables in situ and rapid rheological measurements. This is in direct contrast to established methods, such ascone and plate and Couette rheometry, which are restricted to measurements of liquid viscosity, require large sample volumes, and are incapable of in situ measurements. The proposed technique also overcomes the restrictions of previous measurements that use micro-cantilevers, which are limited to liquid viscosity only, and require independent

measurement of the liquid density. The technique presented here only requires knowledge of the cantilever geometry, its resonant frequency in vacuum, and its linear mass density. A simple yet robust calibration procedure is described to determine the latter two parameters, from a single measurement of the resonant frequency and quality factor of the cantilever in a reference fluid (such as air), if these parameters are unknown.

6. John E. Sader, James W. M. Chon, and Paul Mulvaney (1999) "Calibration of rectangular atomic force microscope cantilevers", Rev. Sci. Instrum. 70, 3967
Abstract: A method to determine the spring constant of a rectangular atomic forcemicroscope cantilever is proposed that relies solely on the measurement of the resonant frequency and quality factor of the cantilever in fluid (typically air), and knowledge of its plan view dimensions. This method gives very good accuracy and improves upon the previous formulation by Sader et al. [Rev. Sci. Instrum. 66, 3789 (1995)] which, unlike the present method, requires knowledge of both the cantilever density and thickness.
7. Nabil Ahmed, Diego F. Nino, and Vincent T. Moy, 2001, "Measurement of solution viscosity by atomic force microscopy" Rev. Sci. Instrum. 72, 2731, (4 pages)
Abstract: We report on studies aimed at employing the atomic force microscope (AFM) to measure the viscosity of aqueous solutions. At ambient temperature, the AFM cantilever undergoes thermal fluctuations that are highly sensitive to the local environment. Here, we present measurements of the (cantilever's resonant frequency in aqueous solutions of glycerol, sucrose, ethanol, sodium chloride, polyethylene glycol, and bovine plasma albumin). The measurements revealed that variations in the resonant frequency of the cantilever in the different solutions are largely dependent on the viscosity of the medium. An application of this technique is to monitor the progression of a chemical reaction where a change in viscosity is expected to occur. An example is demonstrated through monitoring of the hydrolysis of double stranded deoxyribonucleic acid by DNase.
8. P I Oden, G Y Chen, R A Steele, R J Warmack, and T Thundat, 1996 "Viscous drag measurements utilizing micro-fabricated cantilevers", Appl. Phys. Lett. 68, 3814 (3 pages)
Abstract: The influence of viscous drag forces on cantilevers is investigated using standard atomic force microscope (AFM) cantilevers. Viscosity effects on several geometrically different cantilevers manifest themselves as variations in resonance frequencies, quality factors, and cantilever response amplitudes. With this novel measurement, a single cantilever can be used to measure viscosities ranging from $\eta = 10^{-2}$ to 10^2 g/cm.
9. G. Y. Chen1, R. J. Warmack1, T. Thundat1, D. P. Allison1, and A. Huang2, 1994 "Resonance response of scanning force microscopy cantilevers" Rev. Sci. Instrum. 65, 2532
Abstract: A variational method is used to calculate the deflection and the fundamental and harmonic resonance frequencies of commercial V-shaped and rectangular atomic force microscopy cantilevers. The effective mass of V-shaped cantilevers is roughly half that calculated for the equivalent rectangular cantilevers. Damping by environmental gases, including air, nitrogen, argon, and helium, affects the frequency of maximum response and to a much greater degree the quality factor Q. Helium has the lowest viscosity, resulting in the highest Q, and thus provides the best sensitivity in non-contact force microscopy. Damping in liquids is dominated by an increase in effective mass of the cantilever due to an added mass of the liquid being dragged with that cantilever.
10. Abdelhamid Maali1, Cedric Hurth2, Rodolphe Boisgard1, Cedric Jai1, Touria Cohen-Bouhacina1, and Jean-Pierre Aim1, 2005, "Hydrodynamics of oscillating atomic force microscopy cantilevers in viscous fluids" J. Appl. Phys. 97, 074907, 6 pages
Abstract: We present a study of thermal noise of commercially available atomic force microscopy (AFM) cantilevers in air and in water. The purpose of this work is to investigate the oscillation behavior of a clamped AFM microlever in liquids. Up to eight vibration modes are recorded. The experimental results are compared to theoretical predictions from the hydrodynamic functions corresponding to rigid transverse oscillations of an infinitely long rectangular beam. Except for the low-frequency modes, the known hydrodynamic functions cannot describe the amount of dissipated energy due to the liquid motion induced

by the cantilever oscillation. The observed variation of the damping coefficient is smaller than the one predicted. The difference at higher modes between the mentioned theoretical description and experimental results is discussed with the help of numerical solutions of the three-dimensional Navier's Stokes equation.
11. Lippert F, Parker D, Jandt K (2004) Caries Res 38:464
12. Perry J, Neville A, Hodgkiess T (2002) J Therm Spray Technol 11:536
13. Bergauda C, Nicu L (2000) Rev Sci Instrum 71:2487
14. Landau, Lifshitz F (1995) Fluid mechanics, 2nd ed. _Pergamon, New York
15. Sader JE (1998) J Appl Phys 84: 64
16. Sader JE (2000) J Appl Phys 87:3978
17. Boskovic S, Chon JWM, Mulvaney P, Sader JE (2002) J Rheol 46:891
18. Sader JE, Chon JWM, Mulvaney P (2002) Rev Sci Instrum 70
19. Ahmed N, Nino DF, Moy VT (2001) Rev Sci Instrum 72:2731
20. Oden PI, Chen GY, Steele RA, Warmack RJ, Thundat T (1996) Appl Phys Lett 68:3814
21. Chen GY, Warmack RJ, Thundat T, Allison DP, Huang A (1994) Rev Sci Instrum 65:2532
22. Chu WH Southwest Research Institute, Technical Report No. 2, 963
23. Maali A, Hurth C, Boisgard R, Jai C, Cohen-Bouhacina T, Aima J-P (2005) J Appl Phys 97:074907
24. Buenviaje CK, Ge SR, Rafallovich MH, Overney RM (1998) Mater Res Soc Symp Proc 552:187

Chapter 7
Commercial Viscometers

Abstract ***Commercial Viscometers*** for the convenience of the users, availability of various types of viscometers and their manufacturer has been described. Users can choose a viscometer according to their needs. The manufacturers covered are Cambridge Viscosity Inc, PSL Systemtechnik GmbH (Haake viscometer), Anton Paar GmbH, A&D Company Limited Japan, Brookfield Engineering, Cannon Instrument Company. Special viscometers of each manufacturer are described together with their essential features and capability. A list of international manufacturers with their contact addresses and websites have been given. A fairly long list of Indian dealers in viscometers has also been given.

7.1 Introduction

A selected few manufacturers and their products are being described in this chapter. A list, not claimed to be exhaustive, of manufacturers and dealer in viscometers and allied instrumentation is also given towards the end of this chapter.

7.2 Cambridge Viscometers

Cambridge Viscosity, Inc.
101 Station Landing, Medford, MA, 02155, United States
PAC, BP 70285—VERSON
14653 Carpiquet Cedex
FRANCE
www.cambridgeviscosity.com

Cambridge Viscosity Incorporation specialises in fluid viscosity measuring equipment and has world wide offices by different names one such name as obtained from their site is PAC. Their manufacturing ranges in 11 products/services.

S. V. Gupta, *Viscometry for Liquids*, Springer Series in Materials Science 194,
DOI: 10.1007/978-3-319-04858-1_7,

7.2.1 Range of Products

1. Digital viscometers
2. Flow-thru viscometers
3. Laboratory based viscometers
4. Micro-sample viscometers
5. Multi-Channel viscometers
6. Process measuring instruments
7. Process viscometers
8. Sample viscometers
9. Temperature controlled viscometers
10. Viscometers
11. Viscosity measuring equipment.

7.2.2 Viscolab 3000 and Viscopro 8000

One of the specialised viscosity sensing system is their Viscolab 3000, which is very well suited for high-temperature tests where a separate water bath system is not possible. The system contains integrated electronic temperature control. This allows it to deliver accurate measurements at any settings from slightly above ambient to 180 °C. The sensor supports 13 different viscosity ranges. The Viscopro 8000 is designed to measure and control multiple fluid applications in a single production line. Their products are used in oil and gas exploration, refining, biotech, oil analysis and monitoring, fuel combustion, and chemical processing industries.

7.2.3 Various Other Viscometers

They make Capillary, vibration, rolling ball and oscillating piston and electromagnetic viscometers. Basic characteristic of the viscometers are given in Table 7.1.

7.3 HAAKE Viscometer

Germany/International
PSL Systemtechnik GmbH
Pfeiffer System- und Labortechnik
Albrecht-von-Groddeck-Straße 5

Table 7.1 Basic characteristics of viscometers

Viscometer	Capillary	Vibration	Rolling ball	Oscillating piston
Accuracy	± 1 %	N. A	± 1 %	± 1 %
Sample size	25–100 ml	50 ml	500 ml	5 ml
Maximum pressure rating	1500 psi 103 bar	Atmospheric	10 000 psi 689 bar	20 000 psi 1378 bar
Mercury free	Yes	Yes	Yes	Yes
Clean in place	No	Yes	Yes	Yes
Gas viscosity	No	No	No	Yes
Temperature range	Elevated	−40 to 150 °C	Elevated	−20 to 190 °C

38678 Clausthal-Zellerfeld
USA
PSL Systemtechnik
1170 Howell Mill Rd, Suite 300
Atlanta, GA 30318
www.psl-systemtechnik.de/

There appears to be two companies, namely Thermo Electron Corporation Germany and PSL Systemtechnik GmbH having main offices in Germany and USA, which are manufacturing Haake viscometers and instrumentation for reheological measurements.

7.3.1 Range of Products

Main products are: Haake viscotesters 1 plus and 2 plus; Haake viscometers C, D, E; Haake Viscometers 6 plus and 7 plus; Haake falling ball viscometer; Haake viscometer 550, Haake rotovisco 1.

7.3.2 Rotational Viscometers

HAAKE Viscotester 1 plus and 2 plus are battery operated rotational viscometers and are suitable for quick and reliable tests and comparative measurements for quality control applications. The hand-held instruments can also be operated on a stand. These are one button operation instruments. The Viscotester is switched on and off by pushing the button once. Pushing the button again selects the rotor type and starts the measurement.

Contrary to the traditional Viscotester models, where the viscosity value is read from an analogue dial, these viscometers show the viscosity value on a digital display. Therefore, errors caused by misreading the dial are avoidable. Possible handling errors as well as service information are also shown on the display. Units of measurement on 1 plus model is mPa·s and that of VT 2 plus is dPa s.

7.3.2.1 Metrological Specifications

Viscosity range: for 1 plus: 1.5–330 mPa·s and for 2 plus: 0.3–4000 dPa·s
Temperature: up to 150 °C
Rotor speed: 62.5 rpm
Reproducibility: ±1 % FSD
Measuring accuracy: standard: ±5 % FSD optional: ±1 % FSD.

7.3.2.2 Main Features

Main features for Viscotestors 1 plus and 2 plus are:

- Quick, exact and reliable
- One button operation
- LCD display
- No mains supply required
- Typical application fields
- Quick viscosity tests, e.g. for process optimization or machine adjustment
- Batch control in production
- Typical samples
- Printing inks, paints, inks
- Shampoos, creams, lotions
- Oils, greases, pastes
- Sauces, thickeners.

7.3.3 Haake Viscotesters 6 Plus and 7 Plus (Features)

Ready to pack and un-pack; switch on and start measuring viscosity; Digital display of viscosity; % torque; speed, spindle, upper viscosity limit;
Integrated automatic diagnostic functionality;
Visual and acoustic signals at critical measuring conditions;
Digital calibration of the torque with nationally traceable standards;
Digital speed control with built-in accuracy via step motor;
2 years warranty for the most stable measuring instrument in its class 4.

7.3.3.1 Additional Features of the HAAKE Viscotester 7 Plus

Bi-directional RS232 interface
HAAKE RheoWin software (optional) for measurement and data evaluation
Temperature sensor Pt100 to monitor the sample temperature
HAAKE Viscotester 6 plus and 7 plus Standards.

7.3.3.2 Complying Standards

The HAAKE Viscotester 6 plus and 7 plus comply with the following standards:

BS: 6075, 5350
ISO: 2555, 1652
ASTM: 115, 789, 1076, 1084, 1286, 1417, 1439, 1638, 1824, 2196, 2336, 2364, 2393, 2556, 2669, 2849, 2983, 2994, 3232, 3236, 3716.

7.3.4 Falling Ball Viscometer

The HAAKE Falling Ball Viscometer type C provides a very accurate way of measuring the viscosity of transparent Newtonian liquids and gases. It meets the requirements of the German DIN 53015 as well as ISO 12058 standards and it is accepted as an official reference instrument. Its measuring accuracy when supported with the precise temperature control of a circulator is among the highest available in any type of viscometer.

7.3.4.1 Users

1. Chemical industry (polymer solutions, solvents, inks)
2. Pharmaceutical industry (raw materials, glycerine)
3. Food industry (gelatin, sugar solutions)
4. Mineral oil industry (oils, liquid hydrocarbons).

7.3.4.2 Technical Data

- Viscosity range: 0.5–10^5 mPa·s
- Temperature range: −20 to +120 °C
- Reproducibility: <0.5 %
- Comparability: <1 %
- Material: Falling tube, balls 1, 2 and G, Boron silicial glass; balls 3, 4, 5 and 6, Nickel iron alloy.

7.3.4.3 Ball Selection

Selection of ball numbers for various viscosity ranges

Viscosity (mPa·s)	Liquid	Ball number
10^5	Tar	6
10^4	Honey	5 or 6
10^3	Glycerine	4 or 5
10^2	Lubricating oil	3 or 4
10^1	Olive oil or spindle oil	2 or 3
10^0	Water	1
10^{-1}	Ether	1 or G
10^{-2}	Neon	G

7.3.5 Haake MicroVisco 2

The HAAKE MicroVisco 2 unit is a fully automatic, miniaturized version of the traditional falling ball viscometer with integrated Peltier's temperature control. It is specially designed for measurements of optically transparent Newtonian fluids which are of low to medium viscosity and only available in a small sample volume.

7.3.5.1 Applications

Used for precise viscosity measurements of solvent, inks, plasma, serum, beverages, sugar solutions, gelatine solutions, beer.

7.3.5.2 Measurement

The sample to be tested is pulled into the syringe without bubbles and placed into the temperature controlled holder of the HAAKE MicroVisco 2 unit. The sample reached the test temperature after approx. 5 min, and the measuring ball is pulled in the upper start position by means of a magnet. The ball then rolls down the wall of the syringe, which is inclined (15 °C) to avoid bouncing. The fall of the ball is delayed by the resistance of test fluids. The falling ball time is proportional to its viscosity. The test result is the falling time for a defined distance which is measured electronically and converted into viscosity units by the stored calibration factors.

7.3.5.3 Technical Data

- Viscosity: 0.2–1000 mPa·s or in centipoise (cP)
- Repeatability: ±0.3 %
- Comparability: ±0.5 %
- Uncertainty: 0.2–0.7 % (depends on ball used)
- Temperature: 0–100 °C
- Volume: 400 μl
- Measuring time: 3–5 min
- Display in: mPa·s, cP, cSt, ms.

7.3.6 Version "L" or "R"?

Version L viscometers are mainly used for low-viscous substances such as diluted solutions, oils, dispersions and emulsions. Milk-like fluids (2–5 mPa·s) would be an example at the low end of viscosity and motor oil would represent the high end. The measuring range is from 3 to 6000000 mPa·s.

Version R viscometers are used for medium-viscous liquids is the standard viscometer for 80 % of all viscosity measurements. It is especially suited for the medium viscosity range extending from sewing machine oil up to PVC plastisols or whipped cream. The measuring range is from 20 to 400000 mPa·s.

7.3.6.1 Compatibility

The basic ISO 2555 standard describes the design and the characteristic measuring technique of a viscosity measuring instrument (torque, speed, rotor geometry). If a rotational viscometer meets these requirements—as the HAAKE Viscotester 6 plus and 7 plus do—the results comply with the Brookfield method and are 100 % compatible as long as comparable viscometer versions are used.

7.4 Anton Paar Viscometers

Anton Paar GmbH, Anton Paar Strasse-20, 8054 GRAZ, Austria
www.anton-paar.com/viscometers

Anton Paar, besides large number of laboratory instruments and processing instruments, manufactures industrial viscometers and standard oils of known viscosity. The products include Alcohol meter, density meters, CO_2 and oxygen meter, precision thermometers, polarimeter, Rheometer etc. Process instruments are CO_2 sensors, density and sound sensors inline viscometers and polarimeter.

7.4.1 Anton Paar Inline Viscometers

The online viscometer from Anton Paar is integrated directly in the pipe or stirring tank, no bypass is required. The inline viscometer uses a completely new measuring principle based on the measurement of dynamic fluid pressure, which is proportional to a fluid's viscosity. The online viscometer determines viscosities from 1 to 50,000 mPa·s. The viscosity and temperature are displayed as standard. The online viscometer is not affected by drops in pressure or flow rate.

7.4.2 Stabinger Viscometers

The SVM 3000 Stabinger Viscometer measures the dynamic viscosity and density of oils and fuels according to ASTM D7042. From this result, the viscometer automatically calculates the kinematic viscosity and delivers measurement results which are equivalent to ISO 3104 or ASTM D445. The Stabinger measuring principle with Peltier thermostat enables an incomparably wide viscosity and temperature range with a single system. SVM 3000 is quick, compact and energy-saving, versatile in use, needs only small amounts of sample and solvents. It is an efficient high-precision viscometer.

It gives:

- Dynamic viscosity, kinematic viscosity and density in one measurement up to 30 samples/h.
- Automatic filling and cleaning with X sample changers.
- Automatic data processing with VisicoLab software.
- It has high-precision thermostat with stability of 0.005 °C, therefore separate temperature control bath is not required. The measuring temperature can be changed in minutes. It requires small sample size (2.5 ml). Energy consumption is very low (75 W). The software manages calibration constants. Measurement results are independent of gravity and vibration.

7.4.3 Rolling Ball Viscometers Lovis 2000 M/ME

Lovis 2000 M/ME is a rolling-ball viscometer which measures the rolling time of a ball through transparent and opaque liquids and is as per Hoeppler's rolling ball principle. Measurement requires sample of only 0.4 ml. Results are given as relative, kinematic or dynamic viscosity. The instrument is compact and economical, and saves laboratory space. It is the successor to Anton Paar's AMVn automated micro-viscometer.

It can be used for liquids having viscosity of 10000 mPa·s in a temperature range of 5–100 °C.

7.5 A&D Viscometers

A&D COMPANY, LIMITED (Tokyo, JAPAN)
3-23-14 Higashi Ikebukuro, Toshima Ku, Tokyo 170-0013, Japan
www.aand.jp/company

A&D Company, Limited is a world-class producer of advanced measuring, monitoring, controlling and testing instruments. Individuals with varied educational and technical skills use these instruments in a wide range of applications. Yet these products share a number of things in common:

7.5.1 Tuning Fork Vibration Viscometer

A tuning fork vibration viscometer is not the kind of product you hear of very often. This is not surprising since, while there are many viscometers available, A&D is the only company in the world making and selling viscometers based on our unique tuning fork vibration technology. Viscometers in laboratories and research facilities are typically capillary or rotational types. Several companies make vibratory viscometers for production lines, but they use a rotary reciprocating motion with a high frequency of several kHz. Only A&D makes a model that uses a low frequency reciprocating motion like a tuning fork.

7.5.2 A&D Viscometers

Tuning fork sine-wave Vibro Viscometer SV-10/SV-100 measures viscosity by detecting the driving electric current necessary to resonate the two sensor plates at constant frequency of 30 Hz. Viscometers have an accuracy of 1 % of full scale. The range is 0.3–10,000 mPa·s for model SV10. The model SV 100 can measure viscosity up to 100,000 mPa·s. These viscometers have the following advantages:

- Real-time measurement.
- Temperature control: The sensor plate for the SV series is of a thin and small size, in area and in mass, making it is less influenced by the temperature change of the sample fluid and as a result the viscosity measurement tracks the change of viscosity of sample fluid.
- Temperature measurement: It is very important to measure the temperature of the fluid correctly because the viscosity is very much dependent upon the temperature of the fluid. SV series can detect accurate temperature immediately because the fluid and the detection unit (sensor plates) with small surface area/ thermal capacity reach the thermal equilibrium in only a few seconds.
- Continuous measurement: SV series, Tuning-fork Vibration Method does not cause damage to the sample fluid and allows measurement of cloud point of

samples such as surface active agents and of surface/interface changes such as wettability because of its excellent feature for wide measurement range without the need to replace the sensor plates.

- Viscosity calibration: With Viscosity Standard, viscosity calibration can be easily done. 1 point calibration or 2 points calibration is selectable in the calibration mode.
- Non-Newtonian sample viscosity measurement: Thin sensor plates allow little deformation of sample texture and thus enable measurement of stable viscosity values.
- Foaming sample measurement: Low drive frequency of 30 Hz allows measurement of foaming samples without breaking minute foams and with less influence on scattering larger foams.
- Flowing sample measurement: Even the viscosity of flowing samples can be measured, including liquid in turbulent flow, which enables field management with identical data used at the laboratories.
- Separated type model: SV series is composed of separate Display and Main Unit offering excellent placement flexibility.
- Vacuum fluorescent display: You can avoid unnecessary reading errors with easy-to-read, large, clear display with 13 mm height for viscosity measurement and 11 mm height for temperature measurement.
- Small sample size: Standard sample cup requires the sample fluid of just more than 35 ml so you do not need to waste too much sample fluid.
- Easy cleaning: Due to the simple structure, the SUS 304 Stainless steel sensor plates and temperature sensor (all gold plated) and SUS 304 stainless steel protector can be easily and quickly cleaned.
- Sol and gel measurement: Sol and Gel sample fluid like a starch can be measured during the change of material characteristic.
- Standard RS-232C interface: The RS-232C comes standard for your PC or Printer connectivity and the connection cable (25 pin–9 pin) is also standard accessory for your convenience.
- Data collection and graphing software: WinCT-Viscosity (RsVisco) is software to import the measured data of viscosity and temperature to a PC and display graphically the changes in real-time for your analysis.

7.6 Brookfield Viscometers

Brookfield Engineering
11 Commerce Boulvard, Middleboro, MA 02346 USA
www.brookfieldengineering.com
Centurian Scientific Company

CFF-8, Jumbo House Dr Jha Marg, Okhla Industrial Area Phase III, New Delhi, 110020, India www.centurian-scientific.com

Brookfield viscometers are available in a large variety of models with both analogue and digital display. Measurement ranges of viscosity is 0.3 to several million mPa·s.

7.6.1 Rotating Viscometers

Brookfield manufactures both analogue and digital viscometers with a multi-speed electronic drive and ergonomically designed speed with control knobs. Any one of speeds in LVT models may be chosen with a speed knob. This new direct-drive design means extremely quiet operation and greater versatility. The new Universal Power Supply facilitates the use of worldwide power sources. The instruments have continuous torque sensing capability and measurement accuracy of 1 % of full scale range. Repeatability is 0.2 % of full scale range. Easy speed adjustment and on/off control is compatible with all Brookfield accessories. NIST traceable viscosity standards are also available.

7.6.1.1 Features

The basic features of various models are given in Table 7.2.

7.6.2 Falling Ball Viscometers

The viscometers are available in two models namely:

1. **KF30** has a fixed angle of 80° and complies with DIN 53015
2. **KF40** can be angled at 50°, 60°, 70° and 80°.

7.6.2.1 Features

Pivot bearing allows for quick and easy tube rotation, which facilitate repeat tests. There is a connection to circulating bath for temperature control of sample with a temperature probe. To test wide variety of samples a set of six balls is provided. Accuracy ranges from 0.5 to 2.0 % depending on the ball being used. The range of viscosity covered is from 0.5 to 70000 mPa·s. Six balls are provided. The material of ball No. 1 and 2 is Boron silicate glass, of ball no 3 and 4 is Nickel–iron and of 5 and 6 is steel. The diameter varies from 11.1 to 15.81 mm. Length of measurement zone is 100 mm. Temperature range is −60 to 140 °C. Sample required is 40 ml, fall time is from 30 to 300 s.

Table 7.2 Particulars for various Brookfield Viscometers

Model no	Viscosity range mPa·s		Speed in rpm	Number of pre-set speeds
	Minimum	Maximum		
LVT	1	2×10^6	0.3–60	8
RVT	100	8×10^6	0.5–100	10
HAT	200	16×10^6	0.5–100	10
HBT	800	64×10^6	0.5–100	10
DVE Brookfield viscometers with digital display of viscosity % torque spindle speed				
LVDE	1	2×10^6	0.3–100	18
RVDE	100	13×10^6	0.3–100	18
HADV	200	26×10^6	0.3–100	18
HBDV	800	104×10^6	0.3–100	18
DV1 viscometers senses and displays continuously; viscosity in cP or mPa·s, Temperature (°C or °F), % torque, speed (rpm) and spindle used with computer interface				
LVDV-IP	1	2×10^6	0.3–100	18
RVDV-IP	100	13×10^6	0.3–100	18
HADV-IP	200	26×10^6	0.3–100	18
HBDV-IP	800	104×10^6	0.3–100	18

7.6.3 Wells-Brookfield Cone/Plate Viscometers

The Wells-Brookfield Cone/Plate Viscometer/Rheometer is a sophisticated instrument for routinely determination of absolute viscosity of fluids available in small quantities. Its cone and plate geometry provides the development of precise rheological data.

7.6.3.1 Principle

The Cone/Plate Viscometer/Rheometer is a precise torque meter which is driven at discrete rotational speeds. The torque measuring system, which consists of a calibrated beryllium-copper spring connecting the drive mechanism to a rotating cone, senses the resistance to rotation caused by the presence of sample fluid between the cone and a stationary flat plate.

The resistance to the rotation of the cone produces a torque that is proportional to the shear stress in the fluid. This reading is easily converted to absolute units viz. (mPa·s) from pre-calculated range charts. Alternatively, viscosity can be calculated from the known geometric constants of the cone, the rate of rotation, and the stress related torque.

7.6.3.2 Features

The correct relative position of cone and plate is obtained by following a simple mechanical procedure without the need for external gauges. The plate forms the bottom of a sample cup which can be removed, filled with 0.5–2.0 ml of sample fluid (depending on cone in use), and remounted without disturbing the calibration. The sample cup is jacketed and has tube fittings for connection to a constant temperature circulating bath. The viscometer is accurate to within ±1.0 % of full scale range. Reproducibility is within ± 0.2 %. Working temperature range is from 0 to 100 °C.

Various cone spindles are available for use with the Wells-Brookfield Cone/Plate instruments. One cone spindle is provided with the instrument and is calibrated for use with the sample cup. Additional cone spindles may be purchased and will be calibrated for use with the same sample cup.

The Wells-Brookfield Cone/Plate Viscometer/Rheometer provides a wide variety of shear rates and viscosity ranges, which can be further extended by the use of interchangeable cone spindles. Different models can be selected to meet the specific range of viscosities and shear rates required. Shear rates, as high as 1875 s^{-1} (DV-III Ultra Rheometer), can be achieved. The small volume sample requirement permits rheological evaluations to be made on materials where sample availability is limited, such as biological fluids and thick film coatings that contain precious metals. All wetted parts are stainless steel for corrosion resistance and ease of cleaning. Optional purge fitting, lure fitting, and embedded temperature probe are available.

Wells-Brookfield Cone/Plate Digital Viscometers/Rheometers are now supplied with an advanced Electronic Gap Setting feature. This technology allows faster set-up time and minimizes the possibility of gap adjustment errors. With the enhanced system, a bright LED lights up when a hit point is achieved. A simple turn of the micrometer adjustment ring and the gap is set, reducing the time required associated with manual detection methods.

The new electronic gap setting system is now a standard feature of all complete line of Cone/Plate Digital Viscometers/Rheometers. These instruments provide precise shear profiles necessary to determine viscosity and develop rheological data.

7.6.3.3 Data of Cone and Plate Viscometers

The date of cone-plate viscometers is given in Table 7.3.

Table 7.3 Viscosity range in mPa·s for various cone and plate viscometers

Model	Numerals in first row respectively indicate Cone spindle no, sample volume in ml and shear rate (s^{-1}) in N						
	40, 5, 7.5	41, 2, 2	42, 1, 384	51, 5, 384	52, 5, 2	RPM	Increments
LVDV-IIIUCP	0.1–3 k	0.5–11 k	0.2–0.6 k	2–48 k	3–92 k	0.01–250	2.6 k
LVDV-II+PCP	0.2–0.3 k	0.6–11 k	0.3–0.6 k	2–48 k	4–92 k	0.01–200	54
LVDV-IPCP	0.3–1 k	1–3 k	0.6–2 k	5–16 k	9–30 k	0.3–100	18
RVDV-IIIUCP	1–32 k	5–122 k	2–64 k	20–512 k	39–983 k	0.01–250	2.6 k
RVDV-II+PCP	1.6–32 k	6–122 k	3–64 k	25–512 k	49–983 k	0.01–200	54
RVDV-IPCP	3–10 k	12–41 k	6–21 k	51–170 k	98–327 k	0.3–100	18
HADV IIIUCP	2.6–65 k	10–245 k	5–128 k	41–1 M	78–2 M	0.01–250	2.6 k
HADV-II+PCP	3–65 k	12–245 k	6–128 k	51–1 M	98–2 M	0.01–200	54
HADV IPCP	6.6–21 k	24–81 k	12–42 k	102–341 k	196–655 k	0.3–100	18
HBDV IIIUCP	10.5–26 k	39–982 k	20–512 k	163–4 M	314–7.8 M	0.01–250	2.6 k
HBDV -II+PCP	13–26 k	49–982 k	25–512 k	201–4 M	393–7.8 M	0.01–200	54
HBDV IPCP	26–87 k	98–327 k	51–170 k	409–1 M	786–2.6 M	0.3–100	18

7.6.4 High Shear CAP1000+

The CAP 1000+ Viscometer is a compact, heavy-duty instrument used for fast, easy, repetitive testing on the production floor or in the Quality Control Laboratory. Small sample size (of less than 1 ml), along with integrated temperature control built into the sample plate are the plus points. Parallel printer output is also available.

Best results are obtained by following the steps:

1. Specify one speed or shear rate if this is required for your method.
2. Indicate viscosity range.
3. Select the temperature range that best suits your application from 5 to 75 °C or 50 to 235 °C.
4. Choose from 10 spindles for multiple viscosity ranges.

7.6.5 KU-2 Viscometer

The instrument KU-2 is used widely for making viscosity measurements on paints and coatings in accordance with ASTM D562. It is continuous sensing and display in Krebs units, grams, and centipoises (cP) or mPa·s, with rapid change of samples. The resolution of LED digital display of various units is:

1. Krebs Units with resolution of 0.1 KU,
2. grams (in 1 gm increments),
3. centipoise (cP) or mPa·s.

It gives accuracy to ±1 % of full scale range with reproducibility of ±0.5 % of full scale range. The instrument is useful for paints and coatings, adhesives, inks and pastes.

7.6.6 *Process Viscometers*

There are Brookfield rotational viscometers of various models, which provide accurate and repeatable data for on-line viscosity measurement and control. Available models are:

1. AST-100,
2. VTA Pneumatic Viscosel,
3. VTR electric Viscosel.

7.7 Cannon Viscometers

Cannon Instrument Company
2139, High Tech Road, State College PA 16803 USA
www.cannonintrument.com.

7.7.1 *Master Capillary Viscometers*

The Cannon Instrument Company manufactures master viscometers conforming to ASTM D 2162. They also make the following types of glass viscometers as per ASTM D446:

Cannon-Manning BS/IP/MSL, BS/IP/RF U-Tube, BS/IP/SL, BS/IP/SL(S), BS/U; Miniature Tube, BS/U-Tube; Cannon–Fenske Routine, Cannon–Fenske Opaque, Cannon Four-Bulb Shear Dilution, Cannon-Manning Semi-Micro, Cannon-Manning Semi-Micro, Extra Low Charge, Cannon-Manning Vacuum, Cannon-Ubbelohde, Cannon-Ubbelohde Dilution, Cannon-Ubbelohde Semi-Micro, Modified Koppers Vacuum, Ubbelohde, Zeitfuchs Cross-Arm and Zeitfuchs Transparent viscometers.

7.7.2 Viscometers Used in Specific Field

7.7.2.1 Asphalts

Asphalt Institute Vacuum, Cannon-Manning Vacuum, Modified Koppers Vacuum, and Zeitfuchs Cross-Arm.

7.7.2.2 Blood and Plasma

Cannon-Manning semi-micro and cannon-Ubbelohde semi-micro.

7.7.2.3 Low Temperature

Cannon-Ubbelohde.

7.7.2.4 Lubricants and Fuels

Cannon–Fenske Routine, Cannon–Fenske Opaque, Cannon-Ubbelohde and Zeitfuchs Cross-Arm.

7.7.2.5 Paints

Cannon–Fenske Opaque and Zeitfuchs Cross-Arm.

7.7.2.6 Polymers

Cannon Four-Bulb Shear Dilution, Cannon-Ubbelohde and Cannon-Ubbelohde Dilution.

7.7.3 Cannon 2000 Series Viscometers

The CANNON 2000 Series automatic viscometers are bench top versions of the original CANNON Automatic Viscometer (CAV) and are now available in two models namely CAV-2100 and CAV-2200. The CAV-2100 features a single-bath modular construction with two viscometers per bath. The CAV-2200 features a dual-bath modular construction with a single viscometer in each bath. Most other specifications are identical for both models.

7.7.3.1 Time Measurement

The CAV-2000 series viscometers are designed for unattended operation, for the determination of kinematic viscosity testing in accordance with ASTM D 445. The flow time is measured within ±0.001 s by electronically timing the liquid meniscus as it moves between thermistor sensors.

7.7.3.2 Temperature Control

At any selected temperature, bath temperature is controlled with an accuracy of ±0.01 °C within a range of 20–100 °C, and with accuracy better than ±0.03 °C between 100 and 150 °C. Dark or opaque liquids may be measured with the same precision as transparent liquids.

7.7.3.3 Modularity

An outstanding feature of the CANNON CAV 2000 Series viscometers is modularity. As many as four single- or dual-Bath Units may be connected to a single Service Unit and Solvent Dispensing Unit, and the computer.

7.7.3.4 Range of Viscosity

All standard CANNON CAV 2000 Series viscometer tubes accurately measure kinematic viscosity over a 100-fold kinematic viscosity range (e.g., 1–100 mm^2/s, 6–600 mm^2/s, etc.). By selecting two viscometers with complementary viscosity ranges, the range of kinematic viscosity is 0.5–5000 mm^2/s for a single Bath.

Unit Fast-run tubes (10-fold range) are also available for time-critical and/or high-output applications.

7.7.3.5 Sample-Handling

Cannon Instrument Company was the first to provide fully-automatic sample-handling for multiple-sample kinematic viscosity testing. Our pneumatic system is proven and safe with no complicated gears or electric motors and so robust that original CAV instruments over 20 years old are still in use. The Cannon quest for more efficient automatic sample-handling capability has generated a host of high performance options including 10-, 13-, 20- and 50-position sample trays, a Solo Cup test option, plastic and glass sample vials, heated sample trays, and other enhancements. The standard unheated 13-position sample trays allow the operator to load up to 13 samples per viscometer tube for unattended operation. A sample quantity of approximately 12 ml is required for single or double determinations using standard-size sample vials. Throughput for each bath is up to 8 samples/h for each standard viscometer tube, and up to 12–15 samples/h per tube for optional Fast-Run viscometer tubes, depending on sample viscosity and user-configurable test and wash settings. A single instrument can now test up to 100 samples without operator input!

7.7.3.6 VISCPRO® Software

VISCPRO software for Windows 98, 2000, NT, and XP automates multiple samples testing for the CAV. With VISCPRO you can save and restore multiple instrument configurations and/or sample ID information with a few mouse clicks. Three security levels help to protect against inadvertent changes in instrument configuration or sample testing procedures. Analyses can be displayed on the computer screen, printed, saved to an ASCII file or transmitted to the user's Laboratory Information Management System (LIMS) via a serial connection. VISCPRO is shipped with a companion program, the Database Manager, which provides convenient and powerful functions for managing and archiving database data. User-friendly Help files.

7.7.3.7 Safety

The CANNON CAV® 2000 Series incorporates many new safety features, including dual over-temperature fault circuits for each bath. A fluid level float built into the bath prevents the heating elements from operating when they are uncovered. All of the bath fluid expansion from cold-fill to 150 °C is contained within the bath (CAV-2200) or an internal expansion vessel (CAV-2100). An electrical/pneumatic interlock on the solvent waste receiver prevents solvent dispensing with the drain disengaged. The confined sample area and integral vent manifold allows for efficient solvent vapour/fumes extraction. A dual-pane exterior window insulates the warm bath and isolates it from the user. Heated drain lines are now standard on all CANNON CAV® 2000 Series automatic viscometers.

7.7.3.8 Functionality

The CANNON CAV® 2000 Series has two modes of operation: (1) normal (remote) mode, providing fully-automatic computer-controlled operation using Viscpro software and (2) local mode, using instrument panel keypad commands for single drop flow time and kinematic viscosity determinations. Local mode does not require the use of a computer and is particularly useful in diagnostic work. The temperature of the CANNON CAV® 2000 Series bath is monitored by a thin-film RTD sensor and displayed on the front panel. The CAV-2100 includes a "quiet" stirred bath design while the CAV-2200 provides pump circulation of bath fluid;

both systems permit rapid temperature recovery during tube washing and when changing from one test temperature to another. A self-training tube sensor level detection circuit eliminates the need for manual adjustments. Calibration of the CANNON CAV® 2000 Series viscometer tubes is simple—tubes can be calibrated from stored test data without interrupting regular testing. VISCPRO automatically incorporates calibration results into its calculations. Serviceability of the CANNON CAV® 2000 Series viscometers is unparalleled in the industry.

7.7.4 Small Sample Viscometers

7.7.4.1 VISCOlab 3000

The VISCOlab 3000 is ideally suited for high-temperature tests where a separate water bath system is impractical or absent. The range of temperature is from slightly above ambient to 180 °C. With the help of the optional cooling jacket the range goes down to −40 °C. The integrated electronic temperature control allows it to deliver accurate measurements at any setting. The sensor supports 13 different 20:1 viscosity ranges up to 20,000 mPa·s. This provides continuous viscosity, temperature, and temperature-compensated viscosity readings with less than 1.5 ml of sample.

7.7.4.2 VISCOlab 4000

The VISCOlab 4000 supports 12 viscosity ranges, up to 10,000 mPa·s. The required sample size is less than 2 ml of sample. Normal temperature range is from ambient to 110 °C. The optional jacket for temperature control with a re-circulating bath extends the lower temperature to −40 °C. The VISCOlab 4000 features sophisticated electronics that allow users to define temperature/viscosity display units, data averaging interval, and density. It also contains an RS232 interface to make it easy to output data to a PC.

7.7.5 Micro Sample Viscometer

7.7.5.1 VISCOlab 5000

The VISCOlab 5000 requires the smallest sample size of only 75 μl. It was developed for applications where test samples are scarce and very expensive, including drug discovery and biotech. It automatically controls the temperature of the sample from 10 to 40 °C, measures its viscosity, processes the resulting data, and outputs on the integrated screen or to a printer.

7.7.6 *High Pressure Viscometer*

7.7.6.1 VISCOlab PVT

The VISCOlab PVT viscometer can also measure up to pressures of 1379 bar within a temperature range of $-5°$ to $185°$ C. It is designed for high pressure viscosity analysis in oil exploration, core analysis, supercritical fluids, and analysis of dual phase liquid/gas systems. The re-circulating bath requires minimal warm-up time, and includes an intuitive graphical user interface. There is an extra high temperature or extra high pressure option for this system to meet the field requirements of temperature and pressure.

7.7.7 *Process Viscometer Controller*

7.7.7.1 VISCOpro 2000

Featuring menu-driven electronics, the VISCOpro 2000 is the ideal solution for single-channel control needs, offering 13 measurement ranges and memory control for up to 40 fluid settings. The VISCOpro 2000 displays viscosity, temperature, and temperature-compensated viscosity (TCV) parameters, along with alarm signals and control outputs.

7.7.8 *Digital Viscometer*

7.7.8.1 VISCOpro 1600

VISCOpro 1600 is a simple, incredibly flexible, and consistently accurate and can be used as a standalone monitor or a core component of a viscosity measurement system. With multi-channel and LCD display options, the VISCOpro 1600 integrates easily into any PLC. The explosion-proof housing ensures reliability under any conditions.

7.8 List Manufacturers/Dealers of Viscometers

A list of manufacturers and dealers is given in Table 7.4.

Table 7.4 Manufactures and dealers of viscometers

	Manufacturer/Dealer	Viscometers
1	Thermo Scientific Top of Form 2 Radcliff RoadTewksbury, MA 01876, United States	Falling ball haake, Rrotational haake, Process viscometer
2	Kinematica AG Luzerner strasse 147a CH-6014 Luzern Email: info@kinematica.ch	Portable viscometers
3	Galvenic Applied Science Inc. 41 Wellman Street, Lowell, MA 01851, USA Tel: 978-848-2701 Fax: 978-848-2713 Email: info@galvanic.com	Process viscometers
4	IPT www.directindustry.com	Viscometers max 1o l/min, −10 to 65 °C
5	Avenisense, 17, allée du lac Léman Savoie Technolac, BP 233 73374 Le Bourget du Lac CEDEX France	On line fluid analysers for viscosity, density, concentration, temperature, ATEX/DEVIL
6	Measurement	On line fluid analysers for viscosity, density, dielectric constant, temperature, 0.5–50 mPa·s
7	Rheortec Messtechnik GmbH Schutternwaelder Strasse 23 D-01458, Ottendorf-Okrilla Email: info@rheotec.de	Capillary viscometers, falling ball viscometers
8	Narcross Corporation 255, Newton Ville Avenue, Newton MA 02458 www.viscosity.com	In-line viscometers 0.1–7000 mPa·s, 0.1–10000 mPa·s and 100–1000 000 mPa·s
9	PAC BP 70285-VERSON 14653 Carpiquet Cedex FRANCE sales.france@paclp.com Provisco Tech Pvt. Ltd. +91 2227691262, India Email: subhash@proviscotech.com	Automated micro-viscometers, VISCOlab 5000, automated sampling viscometers VISCOlab 3000
10	Hydramotion Ltd, 1 York Road Bussiness Park, Malton York, Y0176YA, England sales@hydramotion.com	Laboratory viscometers, PET and process viscometers
11	Koehler Instruments Company Inc. 1595 Sycamore Avenue Bohemia, New York 11716, USA info@koehlerinstrument.com www.koehlerinstruments.com	Rotational viscometers ASTM D2983, BVS 5000, K447-Mx, K447-SX
12	Ceramic Instruments Via Regina Pacis, 20/26, 41049 Sassuolo MD,e./ C.F. 00710640368	Laboratory viscometers
13	Kittiwake, 3-6 Thorgate Road, Littlehampton, West Sussex, BN17 7LU, UK Email: marketing@kittiwake.com	Heated oil viscometers, falling ball viscometers,
14	Gottfert Material Testing Machines GmbH, Siemen Strasse 2, Postfach 1261, 74722 Buchen, Germany Email: info@goettfert.de	Viscometers

Table 7.5 Manufacturers and dealers of viscometers and related equipment

Aatous International Pvt. Ltd	A-2, Shreeji Society, Plot No. 27, Sector-3, New Panvel, Navi Mumbai-410206, http://www.concretetestingequipments.com/
Aditya Scientific Instruments	1, Agarwal Colony, Civil Lines, Ajmer
Advanced Technocracy Inc.	Atico House No. 5309, Grain Market, Near B D senior Secondary School, Ambala, 133001
Aga group international	Aga House, 173, V.P.O. Dhulkot, Near Sena Nagar, Ambala City-134007, http://www.agalabs.in
Aliallanacollege Of Pharmacy	Molgi Road, Akkalkuwa, Maharashtra
Ambala Scientific Works	Office and Workshop # 2, Rampur Sarsehri, Near Industrial Estate, Jagadhari Road, Ambala Cantt-133001, http://ambalascientificworks.tradeindia.com/
Arunodaaya Export	Sree Gajanan Darshan Chs, E-4/2-3, Sector-7, Sanpada, Navi Mumbai
Associated Scientific & Engineering Works	6, Rani Jhansi Road, Motia Khan, Delhi-110055, http://www.materialtestinginstruments.net
Centurion Scientific	Cff-8, Jumbo House, (Dr. Jha Marg), Okhla Industrial Area, Phase-III, New Delhi-110020, http://www.centurion-scientific.com/
Chandra Scientific Industries	No. 19, Ansari Market, Darya Ganj, New Delhi, Delhi-110002, http://www.indiamart.com/chandra-scientificindus/lab-testing-equipment.html
Classic Scientific	W/32-B, MIDC, Tarapur Industrial Area, Navapur Road, Boisar, Thane-401506, http://www.classicscientific.co.in/
Cole-Parmer India	403-404, Delphi-B, Hiranandani Business Park, Powai, Mumbai-400076, Maharashtra, India☞ Catalogs: http://coleparmer.tradeindia.com/
D.R.F Equipments	76-A, 'C' Block, Dall Mill Compound, Dilshad Garden, Delhi-110095, http://www.drfequipments.com
Dynamic Bio Medical Systems	111, 1st Floor, "Gyan Deep Complex", Laxmi Market, Opp. Canara Bank, New Delhi-110067, http://www.dbmsindia.net/
G. S. Instruments Co.	529, Krishna Mandir Marg, Nangloi, New Delhi-110041, http://www.gsinstruments.net/
Glass Agencies	No. 5309, Anaj Mandi Ambala cantt, Haryana-133001, http://www.indianmart.com/glass-agencies/oil-petroleum-testing-instruments.html
H. L. Scientific Industries	Bank Road, Opp. Sub Power House, Mohan Lal Building, 1st Floor, Ambala Cantt-133001, http://hlscientific.tradeindia.com/

(continued)

Table 7.5 (continued)

Hally Instruments	112, Vasupujya Indl. Estate, Opp. Laxmi Indly Estate, Laxmi Nagar, Goregaon (West), Mumbai-400062, http://hallyinstruments.tradeindia.com/
Hexatec Instruments Pvt. Ltd.	Geeta Udyog Nagar, Bldg. No. 1, Gala No. 5, Gokhiware Village, Nr. Range Office, Valiv Rd, Vasai (E.), Vasai-401208 http://hexatecinstruments.tradeindia.com
Hindustan Apparatus Mfg. Co	114, Favourite Industrial Estate, Masrani Lane, Halav Pool, Kurla (W), Mumbai-400070, http://www.bombcalorimeter.net
Hi-Tech Equipments Inc.	Cb-59, Sector-I, Salt Lake City, Near Pnb-Tank No. 2, Opposite Mafatlal Showroom, Kolkata-700064, http://www.hospiquip.net/
India	Sarigam, Vapi, Gujarat
Industrial & Commercial Services	Plot No: 47, Street No. 11, Near Balaji Temple, Hmt Nagar, Nacharam, Hyderabad-500076 http://ics.tradeindia.com/
Industrial & Commercial Services	Plot No. 47, Street No. 11, H. M. T. Nagar, Nacharam, Hyderabad, Andhra Pradesh-500076, India http://www.indiamart.com/industrial-commercial-services/lab-instruments.html
Jay Instruments & Systems Private Limited	No. E-16, Everest Tardeo Road, Mumbai-400034, http://jayinstruments.tradeindia.com/
Jay Instruments & Systems Pvt. Ltd.	E-16, Everest, Tardeo, Mumbai-400034
Jsr Instruments	711/87, Mathura Vihar, Maktool Puri, Distt. Haridwar, Roorkee-247667, http://jsrinstruments.tradeindia.com/
Kaizen Imperial	B-1/6128, Dev Nagar, Karol Bagh, New Delhi-110005, http://www.kaizenimperial.com/
Lab Science Products	Plot No. 74, Bashyam Street, Devi Nagar, Railway Carshed Complex, Avadi, Chennai-600109, http://labscience.tradeindia.com
Lab Soul, India	No. 20/B, Old Tribune Colony, Ambala, Haryana-133001, http://www.indiamart.com/labsoul-india/industrial-meters.html
Luxmi Instruments & glass Works	14-15B Rani bagh, Ambala cantt
M. G. Scientific Traders	No. 3351, Meena Mal Ka Ahatha, Near Pul Chameli, Ambala Cantt-133001, Haryana, India☞ Catalogs: http://www.scientificinstrumentworld.com/

(continued)

Table 7.5 (continued)

Manish Scientific Instruments Company	No. 83, Jyothi Complex, M.K.N. Road, Guindy, Chennai-600032, (No website)
New Ghosh Industries	71, Kailash Bose Street, Kolkata-700006 http://www.newghoshindustries.net/
New Scientific Instruments	E-2/84, Shivram Park, Nangloi, Delhi
Om Enterprises	D-49, Bhosale Park, Opp. Vaibhav Theater, Hadapsar, Pune-411028, http://omenterprises.tradeindia.com/
Petro-Diesel Instruments Company	Bhakunda Adhar Chanra Adak Lane, Chandan Nagar, Hoogly-712136, http://petrodieselinst.tradeindia.com
Popular Science ApparatusWorkshops Private Limited	3724, Opposite Kali Bari Mandir, Ambala Cant, Ambala, Haryana-133 001 http://www.indiamart.com/popularworks/oil-petroleum-eqiupments.html
Premier Lab Solution	No. 17/19, Anna Nagar, 3rd Street, Behind Surya Hospital, Ganapathy, Coimbatore, Tamil Nadu-641006, http://www.indiamart.com/durapowersolution/thermal-lab-equipments.html
Rico Scientific Industries	Wz-91/B, Tatarpur, Tagore Garden, New Delhi-110027, http://ricoinds.tradeindia.com/
Sams Advanced Climatic Technologies Pvt Ltd (Samsact)	Plot No. 646, Sri Swamy Ayyappa Society, Madhapur, Hyderabad-500081, http://samsadvancedclimatictechnologies.tradeindia.com/
Scientific International	Regd Office: 547, Sba, Plot No. 15, Sector-13, Rohini, Delhi-110085 http://scientificinternational.tradeindia.com/
Scientific International	1590, Madarsa Road, Kashmeri Gate, Delhi-110006 http://www.laboratory-testing-instruments.com/oil-petroleum-testing-instruments.html
Scientific Works Double R. Optics	73, Preet Nagar, Near Raja Park, Jagadhri Road, Ambala Cantt-133001 http://doubleroptics.tradeindia.com/
Shambhavi Impex	No. 108, Neco Chamber, 1st Floor, Plot No. 48, Sector-11, CBD, Belapur, Navi Mumbai, 400614 http://www.shambhaviimpex.com

(continued)

Table 7.5 (continued)

Shivam Instruments	C-14/1, First Floor, Front Side, Wazirpur Industrial Area, Delhi-110052, http://shivaminstruments.tradeindia.com/
Stech Engineers	Unit 4 & 5b, Kiran Indl. Estate, Old Jivdani Road, Opp. Medical College, V.S. Marg, Virar (E), Thane-401303, http://stechengineer.tradeindia.com/ Geeta Udyog Nagar, Bldg. No. 1, Gala No. 5, Gokhiware Village, Nr. Range Office, Valiv Rd, Vasai (E.), Vasai-401208
Sushil Traders	211, The Grand Mall Complex, Opp. Sbi Zonal Office, Nehrunagar, Ambawadi, Ahmedabad-380015, http://sushiltraders.tradeindia.com/
System Anatech (India) Pvt. Ltd.	29-30, 2nd Floor, Mistry Building, 635, J.S.S. Road, Dhobi Talao, Mumbai-400002, http://systemanatech.tradeindia.com/
Teja Scientific Glass Works	7-1-282/C/I/A1, B.K. Guda, Beside Ibp Petrol Pump, Balkampet, Hyderabad-500018, http://www.tejascientificglassworks.com/
Uni-Tech Sales	D-26/11, Gali No. 4-C, Anand Prabat, Industrial Area, New Rohtak Road, New Delhi-110005, Delhi, http://uts.tradeindia.com

7.9 List of Indian Manufacturers and Dealers of Viscometer and Related Equipment

A list of manufacturer/dealers of viscometers and related equipment is given in Table 7.5.

Reference: http://www.tradeindia.com/manufacturers/indianmanufacturers/laboratory-viscometer.html

Chapter 8
Viscosity of Water

Abstract Water is the primary standard of viscosity. Hence measurement of its viscosity has been discussed in detail. The work carried out at NBS (NIST) USA has been described. Swindells capillary viscometer has been described in detail. Swindells measured the viscosity of water by measuring the pressure difference between the ends of a capillary for a known rate of flow. The pioneer works of Roscoe and Bainbridge, Torklep and Oye, Kestin and Shankland and Berstad et al. along with the estimated uncertainty, have been described in detail. Kestin and Shankland arrived at the uncertainty value by taking the sum of component uncertainties and got the value 0.26 %. A small comment about this method has been offered. A shortcoming of not taking into considering uncertainty calculated by type B method has also been pointed out. Overall review by Bauer et al. of PTB has been given. Viscosity temperature relationship has been discussed. Viscosity of various oils used as standards at various temperatures has been given. Dependence of viscosity of water on pressure is quite weak. So only a viscosity coefficient with respect to pressure has been indicated.

8.1 Water as Primary Standard of Viscosity

An accurate knowledge of the viscosities of liquids in absolute units (mPa·s) is of fundamental importance in many scientific fields. To determine viscosity of a liquid even with 1 % uncertainty is a tedious and time consuming job. However lot of laboratory type instruments is available with repeatability of 0.01 %. Hence measuring viscosity relative to the viscosity of known substance is far more easy and accurate. This substance is then taken as primary standard. Considering the easy availability of water of required purity, the water is used as a primary standard of viscosity. The current practice, through out the word, is to use viscosity of water at 20 °C and at one atmospheric pressure as the metrological basis of viscosity measurements. It is, therefore, imperative to discuss measurement of viscosity of water.

S. V. Gupta, *Viscometry for Liquids*, Springer Series in Materials Science 194,
DOI: 10.1007/978-3-319-04858-1_8, © Springer International Publishing Switzerland 2014

8.2 Viscosity of Water

Two types of viscometers have been employed to measure viscosity of water namely (1) Capillary flow meters and (2) oscillating viscometers.

Swindells et al. [1] at NIST, US used capillary flow method. Roscoe and Bainbridge [2], Torklep and Oye [3], Kestine and Shankland [4] and Berstad et al. [5] used viscometers with oscillating, cylinder, disc, cup or the sphere. A pendulum, with a hollow spherical bob with liquid inside it, has been used by Kearsley et al. at NIST (NBS) USA [6, 7] for other liquids. Theory and details of oscillating viscometers have been discussed in the Chap. 5. Flow through a capillary and capillary viscometers have been discussed in Chaps. 1 and 3.

8.3 Viscosity Through Capillary Flow

8.3.1 Capillary Flow Equations

Flow through capillary has been discussed in Chap. 1. Viscosity η has been expressed in terms of, flow rate V/T or Q, dimensions of the capillary (radius a, and length L) and the pressure drop (p) across it as:

$$\eta = \frac{\pi p a^4}{8L} \cdot \frac{T}{V} - \frac{m\rho}{8\pi L} \cdot \frac{V}{T} \tag{8.1}$$

Replacing $\frac{V}{T}$ by Q the flow rate we get

$$\eta = \frac{\pi p a^4}{8QL} - \frac{mQ\rho}{8\pi L} \tag{8.2}$$

Applying end correction as discussed in Chap. 1 (Sect. 1.4.1) [8], the (8.2) becomes

$$\eta = \frac{\pi a^4 p}{8Q(L+na)} - \frac{m\rho Q}{8\pi(L+na)} \tag{8.3}$$

From (8.3), p- the pressure difference between the ends of the flow tube, may be expressed as

$$p = \frac{8Q\eta}{\pi a^4} \cdot (L+na) + \frac{m\rho Q^2}{\pi^2 a^4} \tag{8.4}$$

To calculate viscosity from measurements made on a single capillary, numerical values of m and n are required. By measuring p and Q for a given capillary, p/Q can be plotted against Q, which will be a straight line, thus m can be estimated by using the modified form of (8.4) as given below:

$$\frac{p}{Q} = \frac{8\eta}{\pi a^4} \cdot (L + na) + m \cdot \frac{\rho Q}{\pi^2 a^4} \tag{8.5}$$

To ensure the radial flow (parabolic distribution of velocity across the cross-section of the tube), it is necessary to take longer tube, i.e. making na small in comparison of L but this makes estimation of the value of n more difficult and uncertain. To remove the uncertainty in n, one may measure pressure drop across two tubes of different length but of same diameter. Pressure drops p_l and p_2 across two tubes of different lengths but of same diameter are given by

$$p_1 = \frac{8Q\eta}{\pi a^4} \cdot (L_1 + na) + \frac{m\rho Q^2}{\pi^2 a^4} \tag{8.6}$$

$$p_2 = \frac{8Q\eta}{\pi a^4} \cdot (L_2 + na) + \frac{m\rho Q^2}{\pi^2 a^4} \tag{8.7}$$

Subtracting we get

$$p_1 - p_2 = \frac{8Q\eta}{\pi a^4} \cdot (L_1 - L_2) \tag{8.8}$$

Giving

$$\eta = \frac{\pi a^4 (p_1 - p_2)}{8Q(L_1 - L_2)} \tag{8.9}$$

However it is not practicable to ensure that two lengths of capillary have same diameter, even if they are taken from the same tube. Also, while the flow data may show m and n to be constants for each tube, they may not be the same for both. Consequently, assigning the proper subscripts and proceeding as before, the more general form (8.9) is given by:

$$\eta = \frac{(\pi/8Q)(p_1 a_1^4 - p_2 a_2^4) - (Q\rho/8\pi)(m_1 - m_2)}{(L_1 - L_2) + (n_1 a_1 - n_2 a_2)} \tag{8.10}$$

The second term in the denominator of (8.10) is negligible in comparison of the first term if a_1 and a_2 are not greatly different. Hence simplified form of (8.10) is

$$\eta = \frac{(\pi/8Q)(p_1 a_1^4 - p_2 a_2^4) - (Q\rho/8\pi)(m_1 - m_2)}{(L_1 - L_2)} \tag{8.11}$$

8.3.2 Work of Swindells et al.

Swindells et al. of NBS (NIST USA) [1] used expressions (8.5) and (8.10) for accurate determination of viscosity of water at 20 °C. The apparatus used is shown

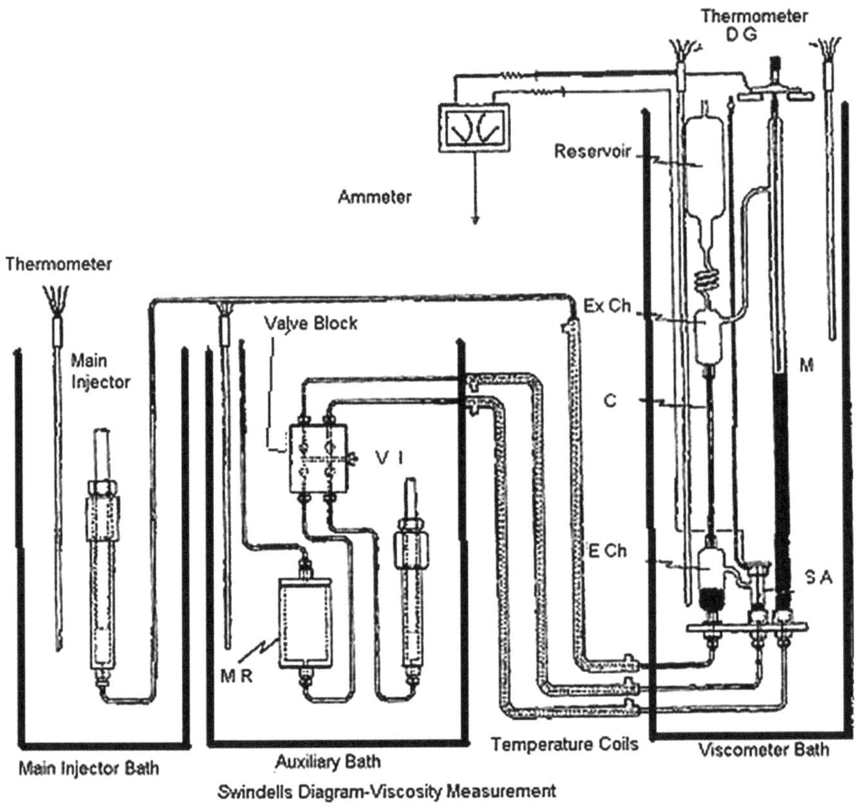

Fig. 8.1 Diagram of Swindle's apparatus for viscosity measurement. *VI* is a valve block, *MR* is a mercury reservoir, *C* is a capillary, *E Ch* is entrance Chamber, *Ex Ch* is exit Chamber, *SA* is the short arm and *M* stands for mercury

diagrammatically in Fig. 8.1. It consisted of three compartments namely (1) Injector, (2) Auxiliary bath and (3) the viscometer.

8.3.2.1 Viscometer

The viscometer consists of a capillary mounted vertically between glass entrance and exit chambers. Glass tubes connect the entrance and exit chambers to the short and long arms of the manometer. A reservoir above the viscometer is connected to the exit chamber by a spiral of glass tubing. Mercury in the entrance chamber and the two arms of the manometer is shown in black colour. The part of the viscometer and manometer above the mercury are filled with water to a level in the reservoir that is higher than the point P where the side tube from the exit chamber connects with the manometer.

The various rates of flow are produced in the capillary by injecting mercury into the entrance chamber at various constant rates, thus displacing water upward through the capillary. The water flows from the exit chamber to the long arm of the manometer and the reservoir, maintaining the same level in each.

Injector

The injector is the arrangement to provide a known and constant rate of flow Q through the viscometer. It consisted of a piston and a cylinder, just like a graduated syringe with out a needle. The piston of the injector moves with the help of a screw, which itself moves by the system of gears and synchronised motor. The injector is calibrated to give the flow rate of mercury in grams per turn of the piston screw at several places of piston along the axis of the cylinder. Swindells et al. measured the flow rate Q with an accuracy better than 1 part in 100,000. The injector was calibrated for delivery rate over a long period of time (1940–1947) and found to be 0.543 92 in 1940, 0.943 92–0.543 96 g of mercury per turn in 1947. The agreement in flow rate per turn is 3 parts per 100,000.

Auxiliary Bath

It consists of, an injector to inject mercury into the differential manometer, mercury reservoir and the valve block. The valve block connects the two arms of the manometer and one arm at a time to the mercury injector and also to each other when required. The temperature inside this bath is kept closed to the temperature of the viscometer bath within 0.02 °C.

Viscometer Bath

The viscometer bath has a viscometer tube, entrance chamber and exit chamber Ex Ch at the ends of the viscometer. The exit chamber Ex Ch ends with a reservoir. The exit chamber is also connected to the longer arm of the differential manometer. The entrance chamber is connected to the injector supplying mercury at the known flow rate. Above the mercury, there is sample water, which also fills, the capillary tube, exit chamber and part of the reservoir. As the longer arm of the manometer and exit chamber is connected with tubing T, the level of water in the longer arm of the manometer is same as in the reservoir.

Manometer

The U tube mercury manometer has two arms, one shorter than the other. The level of mercury in the shorter arm is adjusted to a fixed height. The level in longer arm is adjusted by addition or removal of mercury until the resulting mercury column balances the pressure drops across the capillary. Levels of mercury are measured by measuring the depth of the mercury level from the top of the tube by depth gauges with a micrometer screw head. Top of the meniscus is taken as the reference point in each arm. Electrical contact method is used to measure the depth of mercury level in each arm. Electrical circuit consists of 1.5 V cell, a resistor of 30,000 Ω and a micro-ammeter all in series with the contact. With pointed tips on the rods and reasonably clean mercury-water interface, the mercury does not cling to the point and contact is sharply indicated. The potential of the rods is kept negative with respect to the mercury column. When the rod is kept positive a non-conducting layer of mercury is formed. Even with negative contact, some electrolytic action takes place if the current is left on for too long a time, say a few hours. By this arrangements Swindells et al. are able to measure differential heights with a relative standard deviation of one part in 100,000 (SD of 0.01 mm i.e. 0.0005 inch in the range of 0–40 inch).

For better consistent contacts with the highest point of the mercury meniscus, a disc with a small hole in its centre is fixed at the two points of the arms through which the measuring rod of the depth gauge passes without any lateral movement.

8.3.2.2 Measurement of Pressure Drop

The first step in measuring the differential pressure is to obtain manometer zero, which is the measure of the vertical distance between the contact point in the short arm and the surface plate on which the micrometer rests. With no flow of mercury into the entrance chamber of the viscometer, the level of mercury will be same in each arm. Mercury is added to the manometer system until the equilibrium is reached at the level of contact in the short arm. The level in the long arm is then measured from the surface plate using the longest gage. The micrometer reading for this condition is the measure of the desired vertical distance and is called manometer zero M_O.

To obtain the height of the mercury in the long arm of the manometer for a given flow condition, mercury is forced into the long arm until it stands at the anticipated height. Then with main injector running and water flowing through capillary at the desired rate, the two arms of the manometer are open to each other. Mercury is added or withdrawn from the manometer system as necessary until equilibrium is reached with the mercury in short arm at the level of contact point. The mercury then stands as shown roughly in the Fig. 8.1. The distance from the surface plate is then measured by means of the gage rod of suitable length. The micrometer head reading for this condition is written as Mq. The difference in height h between the levels of mercury in the two arms is then given by

$$h = M_O - M_q + \text{difference in lengths of the two gage rods used}$$

This way, we do not need the actual lengths of each gage rod but only differences in lengths of the particular gage rod from the gage rod used at no flow.

In this system, an increase or decrease in the height of mercury in the short arm of the manometer requires a corresponding flow of water out of or into, the short arm through the side tube connecting it with the entrance chamber. Thus until equilibrium is reached, the rate of flow of water in the capillary is not the same as rate of flow of mercury in the entrance chamber but will certainly correlate with the pressure difference p and hence to the differential mercury height h. If q is the rate of flow through the capillary at that instant and Q is the rate of flow into the entrance chamber. It can be shown that q approaches Q as h approaches the equilibrium value.

Let ρ_w and ρ_{Hg} be density of water and mercury respectively at the given environment. The pressure p is given by

$$p = h(\rho_{Hg} - \rho_w) \tag{8.12}$$

After neglecting the kinetic energy correction term and substituting this value in (8.3), we get

$$h = \frac{8\eta(l + na)}{\pi a^4 (\rho_{Hg} - \rho_w) g} \cdot q \tag{8.13}$$

For brevity writing

$$\frac{8\eta(l + na)}{\pi a^4 (\rho_{Hg} - \rho_w) g} = k \tag{8.14}$$

then

$$q = \frac{h}{k}$$

Till the flow rate q becomes Q, the flow rate into or out of the short arm is given as

$$\frac{1}{2} A \frac{dh}{dt}$$

A is the area of cross-section of the manometer tube and *dh/dt* is the rate of change in pressure head. The total flow rate Q is then

$$Q = \frac{h}{k} + \frac{A}{2}\frac{dh}{dt} \tag{8.15}$$

But (8.15) can be arranged as

$$\frac{dh}{Q - h/k} = \frac{2dt}{A} \tag{8.16}$$

Integrating it we get

$$\log\left(\frac{kQ - h}{C}\right) = -\frac{2t}{kA}$$

$$kQ - h = C \exp\left(-\frac{2t}{kA}\right)$$

or

$$h = kQ - C \exp\left(\frac{-2t}{kA}\right) \tag{8.17}$$

Here C is constant of integration.

Table 8.1 Elapsed time

$\frac{\Delta h_t}{\Delta h_0}$	Elapsed time in minutes			
	2b	2.5b	1A	1.5A
1/2	7	4	10	8
1/5	16	8	25	19
1/10	24	12	55	27

The (8.17) shows that the height of mercury manometer approaches a value corresponding to flow rate Q through the capillary at infinite time.

If Δh_0 be the amount by which h differs from kQ at the start and Δh_t the difference after the elapse of time t, then

$$\frac{\Delta h_0}{\Delta h_t} = \exp\left(\frac{2t}{kA}\right)$$

Giving us

$$t = \frac{kA}{2}\log\left(\frac{\Delta h_0}{\Delta h_t}\right) \tag{8.18}$$

From the knowledge of A and k, one can determine the time to elapse before the readings are taken so that Δh_t is a required fraction of Δh_0. Some typical values of elapsed time are given for the manometer and capillary used by Swindells et al. [1] in Table 8.1.

For such a large time constant, periodic variations in Q resulting from imperfections in the change gears and the injector delivery will not introduce any appreciable errors in the measured pressure.

The measured pressure includes the pressure drop across the tube connecting exit chamber and the longer arm of the manometer. If p is the difference in pressure between the end reservoirs, l_r is the length of the connecting tube, a_r is the radius of the connecting tube, B is the cross-section area of the exit chamber then

$$h(\rho_{Hg} - \rho_w)g = p\left[1 + \frac{l_r}{l}\cdot\frac{a^4}{a_r^4}\left(\frac{A}{A+B}\right)\right] \tag{8.19}$$

Or

$$p = h(\rho_{Hg} - \rho_w)g \cdot F \tag{8.20}$$

F is the correction factor due various connecting links and is given by

$$F = \frac{1}{1 + \frac{l_r}{l}\cdot\frac{a^4}{a_r^4}\left(\frac{A}{A+B}\right)} \tag{8.21}$$

Table 8.2 Correction factor

Capillary	Factor
2.5	0.999995
2.5a	0.999980
1.4	0.999997
1.4a	0.999991

Table 8.3 Designation and length of capillaries

Nominal bore mm	Designation	Length mm	Relative uncertainty
0.5	2.5	48.736	10^{-4}
0.5	2.5a	24.316	10^{-4}
0.4	1.4	45.200	10^{-4}
0.4	1.4a	22.525	10^{-4}

The calculated values of the correction factor for the tubes used by Swindells et al. are given in Table 8.2.

While these corrections are small in comparison to the precision obtained in pressure measurements, the corrections may be applied.

8.3.2.3 Temperature Control in the Viscometer Bath

As the viscosity of water changes by only 1 part in 10,000 for a temperature change of 0.004 °C, the measurement and maintenance of temperature all over the viscometer bath within ±0.001 °C is adequate.

The temperature of main injector bath is kept steady within 0.002 °C. The temperature of the auxiliary bath needs to be maintained only within 0.01 °C. Measurement of temperature is carried out through calibrated platinum resistance thermometers at level of entrance and exit chambers. The difference between the two temperatures should not be more than ±0.002 °C.

8.3.2.4 Capillaries

Swindells et al. used four capillaries of descriptions given in Table 8.3.

Each capillary used for flow measurement is examined by optical methods for (a) circularity (b) uniformity in diameter along its length and (c) inside surface of the tube which may affect viscosity measurement. Swindells et al. measured the diameters of capillary at 45° angular separations for circularity and found the diameters not varying more than 1 % at any point. Average values of the diameter along the length of the tubes were within 5 %. The capillary tubes were slit open along the length and inside surface were examined at 200X magnification. Some scratches of 0.0002 mm were found which were lesser in depth. Inside surface was found to be satisfactory.

The cross-sectional areas of the capillary tubes are measured by gravimetric method with known density of mercury and by measuring electrical resistance with assumed resistivity of mercury.

The effective areas as applicable to fluid flow should be estimated after applying due corrections due to variation in diameters at the particular section and areas along the length.

The square of radius as obtained by the two methods is weighted equally and the mean of the two values is used for each capillary. The effective areas calculated by two methods by Swindells et al. agreed very well. Maximum discrepancy occurred with the capillary 2.5a and that too was only about one part in 10,000.

8.3.2.5 Viscosity Measurement

After assembling the viscometer and manometer, some mercury is injected in the viscometer and both arms of the manometer. Freshly distilled water as specified earlier is introduced into the system through the open tube at the top of the reservoir until desired level is reached. Trapped air is removed by working the main injector back and forth through its stroke. The air trapped in the short arm is forced out through the bleed tube by a forward stroke of the mercury injector. After all the air is removed and the tube is filled with water, the gas tight valve at the top end of the tube is shut off.

The temperature is adjusted in all the baths. The mercury is injected in the short arm of the manometer till the proper height is indicated by the fixed electric contact. The longest of the gauge rods is then fixed in the depth gage head, inserted in the longer arm of the manometer, and the gage head is clamped in position. The gage head is set at its expected reading for the manometer zero and mercury is injected into the longer arm until proper contact is made with the tip of the rod. The two arms of the manometer are now opened to each other. Any change of height of mercury in the longer arm with time is observed and mercury is added or withdrawn in sufficient quantity so that when the final equilibrium is established, the mercury in the short arm makes proper contact with the fixed electric contact. In this situation the levels of mercury in each arm are same. In judging when this equilibrium is reached due consideration is given to the time constant for the particular capillary being used. These adjustments and readings, of course, are made when there is no flow of mercury from the main injector. The head reading of the micrometer for this final equilibrium condition is manometer zero and is designated as M_0.

Manometer heights under flow conditions are measured in a similar manner. The change gears in the main injector are selected to give the desired flow rate. The mercury in the long arm is preset at an expected height, using the proper gage rod. The injection of mercury into the entrance chamber is started and the two arms of the manometer are opened to each other. As in the determination of manometer zero, mercury is added or withdrawn until the mercury level in short arm touches the electric contactor when the equilibrium in pressure corresponding to the particular flow rate is reached. In case the equilibrium is not reached during a single stroke of the injector, the long arm of the manometer is shut off before stopping the main injector. With the manometer still closed the stroke of the injector is reversed. After the forward stroke of the injector is started a second time, the manometer is again open. In this manner the stroke of the injector is repeated as many times as required to establish the equilibrium. The reading of the micrometer head in the equilibrium condition is designated as M_q. This process is termed as one run and a run number is assigned to it. Usually three to four such runs together with repeat observations of the manometer zero can be carried out in a day.

While measurements are being made, a continuous check on the temperature of the viscometer bath and record of the thermometer readings is kept every 5 min. From these records the mean temperature in the viscometer bath is estimated and is used for the particular run. A record of temperatures of the other baths is also kept to ensure their proper working.

The (8.5) may be written as;

$$\frac{p}{Q} = A + B \cdot Q \tag{8.22}$$

where

$$A = \frac{8\eta(l + na)}{\pi a^4}$$

$$B = \frac{m\rho}{\pi^2 a^4} \tag{8.23}$$

The values of A and B are calculated by the least square method using the data of pressure drops for a particular flow rate. Four values of flow rate Q were taken for each capillary. Measurement of Q is much accurate so the values of Q are taken error free. Only errors adjusted by least square method are taken in the pressure drop-p.

8.3.2.6 Measurements Results

Inserting the measured values of radius and length of the capillary and the known value of density of water in A and B, the viscosity values of the water at 20 °C have been calculated. The value of n was taken as 0.57.

The values of viscosity of water at 20 °C were calculated for each capillary using least square method (8.22). Entrance ends were reversed for each capillary. Further viscosity of water was also calculated by pairing the four capillaries two at a time and measuring pressure drops across each capillary. The results are summarised in Tables 8.4 and 8.5 as follows:

Table 8.4 Mean viscosity at 20 °C for individual capillary

Capillary no	2.5	2.5a	1.4	1.4a
Viscosity values at 20 °C	1.0017	1.0012	1.0015	1.0015
	1.0018	1.0015	1.0014	1.0015
	1.0021	1.0010	1.0015	–

Table 8.5 Mean viscosity at 20 °C for capillaries in pairs

Capillary pairs	2.5 and 2.5a	1.4 and 1.4a	2.5 and 1.4a	2.5a and 1.4
Viscosity values	1.0022	1.0014	1.0022	1.0018

The values of viscosity calculated from (8.22) obtained by treating the data of each capillary individually show a spread from 1.0012 to 1.0019 with a mean value of 1.0015. The mean value of viscosity obtained from the combinations of two tubes is 1.0019. This indicates that the value of n taken as 0.57 is too high for calculation of viscosity values for individual capillaries. By taking the same data and assuming n equal to zero, the mean, for all four capillaries, comes out to be 1.0019 which is same as is obtained with the capillaries in pairs. Although the assumption, that n is zero, results in better over-all agreement in the calculated values, the measurements are so insensitive for the evaluation of n that no accurate conclusions may be reached as to its magnitude. In consideration of this uncertainty in the value of n, the best result obtained from the data is

$$\eta_{20} = 1.0019 \pm 0.0003 \text{ mPa}\cdot\text{s}$$

8.4 Viscosity by Oscillating Viscometers

8.4.1 Work of Roscoe and Bainbridge

The main features of the viscometer used by Roscoe and Bainbridge [2] have already been given in Chap. 5 (Sect. 5.5). The formula used was

$$I\delta = \frac{2}{3}R^4(\pi\eta\rho T)^{1/2}\, F(\delta, p) \tag{8.24}$$

The parameters to be measured are I the inertia of the rings, δ logarithmic decrement, R the radius of the bulb, ρ density of water and T the time period of the oscillation. Stability of temperature is an important environmental parameter. The p in the function $F(\delta, p)$ is given as

$$P = R\left\{\frac{\pi\rho}{\eta T}\right\}^{1/2}. \tag{8.25}$$

8.4.1.1 Inertia of the Rings

The two inertia rings were made from same piece of brass. The height of the one ring was twice of the other. The care was taken to avoid strain when the two rings were separated and their ends were machined. The two rings were always used together. The dimensions of the two rings were measured at 20 °C and repeated after 1 month to ensure that no dimensional changes have taken place due to machining. The diameters in different orientations were within 10 μm and variation in the height as 40 μm. The mean internal and external diameters were measured with an accuracy of 2 μm. The mass of the rings was measured and moment of Inertia I was calculated at the temperature of measurement. The moment of inertia I was found to be 29992.51 ± 0.13 g cm^2 at 19.75 °C.

8.4.1.2 Temperature Measurement and Maintenance

The stem of water filled spherical bulb was graduated in such a way (Fig. 5.3) that the water thermometer so formed had a sensitivity of 0.05 °C per mm at 20 °C. The stem was calibrated against calibrated thermometers and the smooth curve drawn was accurate within ±0.002 °C. Standard deviation of individual observed temperature was 0.005 °C. The evaporation of water was checked by occasional weighing the bulb against a fixed set of weights with a detectability of 0.02 mg.

During the measurements of δ (logarithmic decrement), air temperature was kept within 0.05 °C and temperature of water in the bulb within 0.01 °C.

8.4.1.3 Diameter of the Bulb

The diameters of the spherical bulb were differing by only 0.01 mm and mean value of the external diameter was found to be 65.328 mm with a standard error of 0.007 mm. In order to find internal diameter, the apparent thickness of glass in the equatorial region was measured with a travelling microscope in sodium light and was found to be 0.834 ± 0.003 mm. The true thickness was calculated by expression

$$\frac{\mu d}{[1 - (\mu - 1)d/R]} \tag{8.26}$$

where d is apparent thickness, μ is refractive index of glass in Sodium light and R is the external radius of the bulb.

The μ was determined by immersing the bulb in Benzene and adding paraffin oil until the surface of the bulb most nearly disappeared when viewed in sodium light and then measuring the refractive index, which found to be 1.476. The true thickness of glass was thus obtained as 1.217 ± 0.004 mm giving the internal diameter as 62.894 ± 0.011 mm. The effective radius of the bulb R was also calculated by measuring mass of water filled up to different heights of the bulb and the weighted mean of all measurements was found to be 31.4165 ± 0.000 9 mm.

8.4.1.4 Measurement of Decrement

The rotation of the bulb was observed by means of the image of a fine thread formed by lens and reflected by the plane mirror M on to a suitably curved scale placed 3 m away. The bulb was set to oscillate by turning the torsion head through 90°. The observations were started about 10 min later. In each run 64 successive observations were taken of the maximum deflections on each side of zero. The deflections varied from 45 to 1.4 cm. Pairs of left and right deflections were averaged, each average being denoted by A_n where n ranged from 0 to 63. Nine such runs were made.

If A_n the amplitude of the nth observation then A_n can be expressed as

$$A_n = A_0 \exp(-n\delta) + e_n \tag{8.27}$$

The e_n represents error of measurement. The least square method is used to minimize the errors of measurement and to obtain best values of δ.

If E is sum of squares of errors e_n the E is expressed as

$$E = \sum_{n=0}^{n=63} \{A_n - A_0 \exp(-n\delta)\}^2 \tag{8.28}$$

To minimize E, differentiate it with respect to A_0 and put it equal to zero, giving us

$$A_0 = \frac{\sum_{n=0}^{n=63} A_n \exp(-n\delta)}{\sum_{n=0}^{n=63} \exp(-2n\delta)} \tag{8.29}$$

Similarly differentiating E with respect to δ and putting to zero we get

$$A_0 - \frac{\sum_{n=0}^{n=63} nA_n \exp(-n\delta)}{\sum_{n=0}^{n=63} n \exp(-2n\delta)} \tag{8.30}$$

In order to get consistent values of A_0 from (8.29) and (8.30) iteration method is used. Say take an approximate value of δ (calculated from a pair of values of A_n) and calculate the values of A_0 from each equation, then decrease the value of δ slightly and see if the difference in two values of A_0 decreases or increases if decreases, then continue to decrease the values of δ in very small steps, till the difference in two values of A_0 is within the desired accuracy.

Table 8.6 Measurements Results of decrement and Time period

Parameter	With water		With air	
	With ring	Without ring	With ring	Without ring
$2\pi\delta$ Decrement	0.055 280	0.24871	0.001826	0.005859
T Time period	9.0200	5.1149	10.0636	6.3055
Temperature	19.750	20.091	21.5	19.6
Pressure	73.48	72.48	76.4	76.4

Roscoe and Bainbridge obtained the value of A_0 as 44.9708 for first 64 observations and A_0 *as* 25.8531 for last 54 observations i.e. when first 10 observations were omitted. Similarly the values of $2\pi\delta = \Delta$ were 0.0553154 and 0.055 280 2 with standard deviation of 0.0074. Standard deviation, in this case, is the square root of the sum of residual errors divided by 2 less than the number of observations. Further calculations, in which more number of earlier oscillations is omitted, showed no further improvement in standard deviation. Moreover δ was found to be constant.

8.4.1.5 Measurement of Time Period

Time period was determined by measuring time intervals with a calibrated stop watch. A smoothed calibration curve was plotted of error against time interval correct to about 0.01 s. The standard deviation of individual measurements was 0.1 s. The time period was found to be more with larger amplitudes. The time period was found to be 0.2 % higher when the initial amplitude was 100 cm on scale but only 0.002 % higher when the amplitude was 10 cm

Having obtained the values of δ and T, R the viscosity of the water was calculated from (8.24). The function $F(\delta, p)$ is given by Roscoe in series form in (8.11) of Part I [9]. In the paper Roscoe has also discussed the case of an axially symmetrical vessel which departs slightly from the spherical form.

The function F approaches to unity as δ becomes small and p approaches to a large value.

The observations have been taken when the bulb was filled with water with and without the inertia ring. Similarly the observations were taken when the bulb was containing air only with or without the inertia ring. The results are summarized in the Table 8.6.

From row for decrement, we observe that the air resistance contributes to about 3 % to the damping made with the ring. So it is necessary to determine appropriate correction to fairly high degree of precision.

For this purpose a value of moment of inertia I_0 of the suspended part only is also required. In order to calculate I_0, experimental data of the suspended system in water with the ring and without it have been used in (8.31). Air damping adds to the right-hand side of which should approximate to $I\Delta$ calculated from the data obtained for air. Effective decrements Δ_l–Δ_2 should therefore be the differences between decrements in water minus the decrements in air. i.e.

$$\Delta_1 = 0.24871 - 0.00559 = 0.24312 \text{ without ring}$$
$$\Delta_2 = 0.055280 - 0.001381 = 0.053899 \text{ with ring}$$

Equation (8.24) gives

$$\frac{I_0}{I_o + I_1} = \frac{0.053899}{0.24312}\left\{\frac{\eta_1\rho_1 T_1}{\eta_2\rho_2 T_2}\right\}^{1/2}\frac{F(\delta_1,p_1)}{F(\delta_2 p_2)} \tag{8.31}$$

Here η_1 is equal η_2 and ρ_1 is equal to ρ_2, $T_1 = 5.1145$ s and $T_2 = 9.0200$ s and functions $F(\delta_{1,} p_1)$ and $F(\delta_{1,} p_1)$ may approximately be taken as unity. The calculations give the value of I_0 as 574 g cm^2. Actually the air damping depends upon somewhat on the time period also, which is not quite the same in corresponding measurements with and without water. Time periods of pendulum with the rings are 9.0200 and 10.0636 s. Similarly time periods are 5.1149 and 6.3055 without ring. However the value of I_0 obtained by this method is sufficiently accurate for the calculation of air resistance. Once I_0 is known the value of $I_0\delta_0$ is calculated by putting the values of viscosity η and density ρ for air on the right hand side of (8.29).

They have used the following data at pressure of 76 mm Hg and temperature 20 °C:

Air	
Density	0.001 205 g cm^{-3}
Viscosity	0.01820 mPa·s
Temperature coefficient of viscosity	0.00005 mPa·s per °C
Water	
Density	0.99820 g cm^{-3}
Temperature coefficient of viscosity	0.002485 mPa·s per °C
Temperature coefficient of density	0.000 20 g cm^{-3} per °C
Coefficient of expansion of brass ring	2.1 × 10^{-5} per °C
Coefficient of linear expansion of Pyrex glass	0.3 × 10^{-5} per °C
Moment of inertia of the ring I_1	2992.51 g cm^2 (10^{-7} k gm^2)

After applying corrections for smaller effects of air resistance and temperature of measurement which was 19.75 °C, Roscoe and Bainbridge obtained a value of the viscosity of water at 20 °C as **1.0025 ± 0.0005 mPa·s**.

8.4.1.6 Uncertainty Budget

Uncertainty budget is given in Table 8.7.

Table 8.7 Uncertainty budget

Source of error	Contribution to viscosity mPa·s
Decrement measurements	±0.000 33
Mean radius of the bulb	±0.000 25
Moment of Inertia of pendulum alone	±0.000 25
Air damping correction	±0.000 22
Moment of inertia of the ring	±0.000 08
Temperature measurement	±0.000 07
Time period measurement	±0.000 04
Sum of squares of contributions	2952×10^{-10}
Standard uncertainty	0.00054 mPa·s
Relative standard uncertainty	0.054 %

8.4.2 Work of Torklep and Oye

Basic viscometer has already been described in the Chap. 5 (Sect. 5.6). The pendulum and the solid cylinder were checked for straightness with the help of lathe and microscope. Special attention was paid to the flanges. The whole oscillating system was thus set to a linearity of 0.02–0.03 mm or better.

8.4.2.1 Dimensions of the Oscillating Cylinder

The oscillating cylinder was made of Platinum Iridium (90 % Platinum). Initial dimensions were measured and cylinder was heated to 1150 °C. Re-machining became necessary after heating the cylinder to such a high temperature. The dimensions are taken again and a check of dimensions after the experiment showed no change in cylinder's dimensions. In addition roundness diagrams, at the different sections the cylinder, are recorded at the amplification factor of 10,000. The microscopic examinations and the recorded diagrams showed that radius of the cylinder was the same within 1 μm.

Dimensions of the cylinder at different occasions are shown in the Table 8.8.

Two thin rings about 50 mm apart surrounded the oscillating cylinder with a clearance of 0.3 mm. The rings serve to determine the lateral position of the pendulum. Centring is achieved by adjusting the entire viscometer stand using the base plate.

8.4.2.2 Moment of Inertia

The moment of inertia is determined by measuring the time periods of the oscillating system with and without the rings of known dimensions and mass. The standard deviation of measurement of moment of inertia is 1.5×10^{-9} k gm^2.

Table 8.8 Dimensions of the cylinder at different occasions

Sections	Before heating mm	After machining mm	Later occasions mm
Top	18.924	18.451	18.455
	18.924	18.454	18.457
	18.924	18.455	18.461
Middle	18.924	18.452	18.454
	18.925	18.456	18.454
	18.923	18.456	18.458
Bottom	18.924	18.453	
	18.925	18.455	
	18.924	18.454	
Mean value	18.924	18.544	18.454

Moment of inertia of oscillating system and rings is a temperature dependent quantity and therefore, is calculated very carefully at the temperature of measurement. The following relation was used.

$$I_1 = I_2 \frac{T_1^2}{T_2^2} \tag{8.32}$$

Here I_1 and I_2 are the moments of inertia of the same body at temperatures T_1 and T_2 respectively.

8.4.2.3 Measurement of δ and T

Logarithmic damping δ and time period T in water and air were measured by measuring the time interval required to travel the distance between the two fixed diodes for several consecutive excursions of the oscillating system. The two diodes were kept 100 mm apart and roughly symmetrical to the oscillation zero of the reflected beam. Laser beam sent through a lens system of one convex and two cylindrical lenses to form a line-shaped image was reflected from the mirror on the pendulum to fall upon two photodiodes. The distance between the planes of photodiodes and pendulum was kept 2 m. The temperature was allowed to reach equilibrium over night. The damping constant and time period were measured in air for 50 oscillations after omitting first 10 oscillations. Total 10 such runs were repeated. Temperature controlled container was thereafter filled with water to a height of 15 mm above the oscillating cylinder. Three series of control measurements were carried out on the distilled water passed through a high grade ion exchange column. The temperature was allowed to stabilise over a period of several hours. Air bubbles adhering to the cylinder were removed. In this case also 10 runs of 50 consecutive oscillations were taken.

The period and damping constant both increase slightly with amplitude. The coefficients of increase are respectively 0.12 ms/m and 5×10^{-5}/m. These values do not depend significantly on the nature of braking media or the magnitude of the

Table 8.9 Viscosity of water

Temperature	Viscosity mPa·s	SD 10^{-4} mPa·s	ICPS 1974 mPa·s
24.931	0.8919	3	0.8916
24.951	0.8919	4	0.8912
24.938	0.8923	4	0.8915

damping constant. The amplitude effects are therefore due to non ideality in the torsion wire response (torsion constant or initial friction). It has been seen that 8 % Pt alloy wire is much better than a tungsten wire. The Pt–W wire has a coefficient of damping of 3 % and time period as 5 % respectively of those for tungsten wire of similar dimensions.

8.4.2.4 Data Used

Density of air at temperature T in kelvin = $0.3532 \times 10^3/T$ kg/m^3

Viscosity of air η_t at temperature T in kelvin is taken as

$$\eta_t = 2.843 + \left(0.80 - 1.47 \cdot 10^{-4}T + 4 \cdot 10^{-8}T^2\right)(\log(T/273)\ \mu\text{Pa}\cdot\text{s}$$

Density ρ of water at temperature t °C

$$\rho = \left[0.997296 - 2.56 \cdot 10^{-4}(t - 24)\right] \cdot 10^3\ \text{kgm}^{-3}$$

Note: log here means logarithm to the base e.

8.4.2.5 Results of Measurement

Results from three series of experiments of water viscosity carried out at intervals of 3–4 months are given in Table 8.9. The mean value of each series and the standard deviation of the mean are listed.

8.4.2.6 Uncertainty Budget

The last row of Table 8.10 indicates that combined standard deviation obtained by using damping (8.15) of the Chap. 5 is much smaller than that obtained by using time period (8.16). The (8.16) of Chap. 5 is extremely sensitive to the term $(1 - 1/\omega^2)$. The term, even with a very small uncertainty in ω, affects vastly the value of viscosity η calculated from the (8.16). Hence most of the time (8.15) of Chap. 5 is used.

Table 8.10 Uncertainty budget

Parameter	Estimated standard deviation		Relative viscosity change	
	Absolute	Relative	By damping	By period
Radius	1 μm	0.0001	−0.0007	−0.0007
Damping constant	5×10^{-6}	0.0003	0.0006	≈0
Moment of inertia	1.5×10^{-9} k gm^2	0.0002	0.0003	0.000 3
Height	2 μm	0.0001	0.0002	−0.0002
Temperature	0.01	–	0.0002	−0.0002
Rod radius	50 μm	0.07	0.0002	≈0
Rod depth	1 mm	0.07	0.0001	≈0
Period in air	20 μs	0.000 01	≈0	−0.01
Period in water	20 μs	0.000 01	≈0	0.01
Combined			0.00103	0.014

8.4.3 Work of -ShanklandKestin and Shankland

8.4.3.1 Viscosity Equations

In their earlier viscometer, the disc used to oscillate between two fixed plates to counteract the effects of secondary flows and to compensate for differential thermal expansion. Kestin and Shankland [4] developed a viscometer in which there were no fixed plates. They also developed the necessary mathematical expressions to calculate viscosity by measuring damping constant and time period of the free oscillating disc.

Their working equations are as follows:

$$\frac{2}{\omega}\left(\frac{\delta}{\omega} - \delta_0\right) = g_0 \rho y \left[g_1 H_1 + g_2 y/\omega + g_3 y^2/2g_4\omega + y D_R(\eta_0)/R\omega\right] \quad (8.33)$$

and

$$\left(\frac{\delta}{\omega} - \delta_0\right)^2 + 1 - 1/\omega^2 = g_0 \rho y \left[g_1 H_2 + g_2 y\delta/\omega - g_3 y^2 g_4 - y Q_I(\eta_0)/R\omega\right] \quad (8.34)$$

Here

$$y = \left(\frac{\eta\, T_0}{\rho\, 2\pi}\right)^{1/2}, \ \omega = T/T_0, \ g_0 = \pi R^4/I \quad (8.35)$$

$$g_1 = 1 + 2d/R, \quad g_2 = \left[\frac{16}{3\pi}\left(\frac{4\pi}{3\sqrt{3}} - 1\right) + \frac{3d}{R}\right]/R, \quad g_3 = \left(\frac{17}{9} + \frac{3d}{4R}\right)/R^2$$

$$g_4 = \left[\frac{(\delta^2+1)^{1/2} - \delta}{2\omega}\right]^{1/2}, \quad H_1 = \frac{3g_4}{2\omega} - \frac{1}{8g_4^3\omega^3} \text{ and } H_2 = \frac{3}{4g_4\omega^2} - g_4^3$$

$Q_R(\eta_0)$ and $Q_I(\eta_0)$ denote the real and imaginary parts of $Q(\eta_0 i^{1/2})$ respectively.

In fact (8.15) and (8.16) by Berstad et al. [5] given in Chap. 5, are the modified forms of (8.33) and (8.34).

8.4.3.2 Viscometer

The design and construction of the viscometer has been described in [10] and fully described in Chap. 5 (Sect. 5.6). For measurement of water viscosity, they made only one change, namely they removed the fixed plates guarding the oscillating plate. The reason was that in case of water boundary layer thickness was about one tenth of the distance between the oscillating disc and free stationary surface of water.

For the purpose of estimating the standard uncertainty, neglecting g_2, g_3, Q_R and δ_0 in (8.36) they have roughly expressed η as:

$$\eta \approx \frac{16\pi}{g_0^2 g_1^2 \rho T_0}\left(\frac{\delta^2}{\omega}\right) \tag{8.36}$$

This equation represents 85 % of the measured viscosity and is sufficiently accurate for the purpose of estimating uncertainty in terms of uncertainties of the input quantities.

Taking log of (8.36)

$$\log \eta = \log 16\pi + 2\log\delta - 2\log g_0 - 2\log g_1 - \log\rho - \log T_0 - \log\omega \tag{8.37}$$

Differentiating we get:

$$\frac{\delta\eta}{\eta} = 2\frac{\delta(\delta)}{\delta} - 2\frac{\delta g_0}{g_0} - 2\frac{\delta g_1}{g_1} - \frac{\delta\omega}{\omega} - \frac{\delta\rho}{\rho} - \frac{\delta T_0}{T_o} \tag{8.38}$$

By applying the law of relative variances, (8.39) gives

$$v_s^2 = 4v_\delta^2 + 4v_{g0}^2 + 4v_{g1}^2 + v_\omega^2 + v_\rho^2 + v_{T0}^2 \tag{8.39}$$

Here v stands for relative variance and sub-script is for the parameter contributing to the uncertainty.

We know that relative uncertainty due to a parameter is the square root of the relative variance of that parameter. Hence expression (8.40) may be rewritten in terms of relative uncertainties as

$$u_s^2 = 4u_\delta^2 + 4u_{g0}^2 + 4u_{g1}^2 + u_\omega^2 + u_\rho^2 + u_{T0}^2 \tag{8.40}$$

Here u stands for relative uncertainty and sub-script is for the parameter contributing to the uncertainty.

Table 8.11 Uncertainty budget

Parameter	Relative uncertainty in %
δ	0.05
ω	0.0.01
g_0	0.04
g_1	0.0.02
ρ	<0.005
T_0	0.03

8.4.3.3 Uncertainty Budget

The parameter wise values of component uncertainties as obtained by them are given in Table 8.11.

Kestin and Shankland [4] took the relative standard uncertainty as the sum of the absolute relative uncertainties of the components affecting the measurements and gave the relative uncertainty of their measurement as 0.26 %.

However the standard uncertainty u_s should be calculated from (8.40).

From the data given in Table 8.11, the value of relative standard uncertainty will then be 0.13 % only. This is exactly half of that quoted by Kestin and Shankland [4].

8.4.3.4 Viscosity and Temperature Relation

The viscosity of water was measured by Kestin and Shankland [4] in the temperature range of 25–150 °C. From the data obtained, the viscosity at t °C in the temperature range of −8 to 150 °C, has been expressed in the form of two empirical relations:

$$\log\left(\frac{\eta(t)}{\eta(20\,^{\circ}\mathrm{C})}\right) = \frac{(20-t)}{(96+t)}\sum_{i=0}^{i=3} A_i(20-t)^i \tag{8.41}$$

Two sets of values of A_i are given Table 8.12. The A_i, for the temperature range of 0–40 °C are given in column 2 of Table 8.12. For much larger range of −8 to 150 °C the A_i, are given in column 3 of the Table 8.12. Here η is dynamic viscosity and t is temperature in °C. Associated uncertainty in $\eta(t)/\eta(20\ ^{\circ}C)$ is 0.05 % in the temperature range of 0–40 °C but in wider range of −8 to 150 °C this uncertainty increases to 0.2 %.

Table 8.12 Values A_i

	0 °C ≤ t ≤ 40 °C	−8 °C ≤ t ≤ 150 °C
A_0	1.2364	1.23788
A_1	-1.37×10^{-3}	1.303×10^{-3}
A_2	5.7×10^{-6}	13.06×10^{-6}
A_3	0	2.55×10^{-8}

8.4.4 Work of Berstad et al.

8.4.4.1 Theoretical Considerations

For the validity of (8.15) and (8.16) of Chap. 5, three basic requirements are:

1. The cylinder radius R is large compared to the thickness of the boundary layer. Thickness of the boundary layer y is expressed as

$$y = \left(\frac{\eta}{\rho}\frac{T}{2\pi}\right)^{1/2} \tag{8.42}$$

   In case of Berstad et al. y was 0.5 mm. So the ratio of the cylinder radius to boundary layer was about 20 and 17 for the two cylinders used. Radii of two cylinders were 10 and 8.5 mm.
   In (8.15) and (8.16) of Chap. 5, $x = R(2\pi\rho/\eta T)^{1/2}$ is equal to the radius divided by the boundary layer thickness y (8.35A) and therefore is a pure number around 20.
2. The cylinder should be thick. That is, the ratio of half height to the boundary layer thickness is of the order of unity or larger this value was 20 and 34 for two cylinder used.
3. The oscillating body should be free i.e. the surrounding liquid is at rest at a distance large compared to the boundary layer thickness. The nearest object, including the water–air interface from the oscillating cylinder or sphere should be at least 10 mm. This ratio was 20 in Berstad experiment.

8.4.4.2 Sample Water

The water used in the viscosity measurement was deionised and then distilled in 4 stage still of Pyrex glass. The water was further purified by through reagent grade chemicals. The filtration ensured the removal of all organic compounds and micro-organisms larger than 0.22 μm. However no isotopic composition was checked. Though undue exposure to atmospheric air was avoided, but no dissolved air was removed before introduction in the viscometer chamber. The water used was transported in closed polypropylene bottles. No air bubbles were found to be adhered to the oscillation body.

Table 8.13 Particulars of oscillating bodies

Bodies	Cylinder 1	Cylinder 2	Sphere
Material	Stainless steel ASI304	Stainless steel ASI304	SIS 2258 steel
Radius (mm)	9.9862	8.5046	11.9057
Height (mm)	20.0055	33.9950	–
Dimensional accuracy	±0.25 μm 99 % CL	±0.25 μm 99 % CL	±0.25 μm
Roundness	1.5 μm	1.0 μm	0.8 μm

8.4.4.3 Particulars of Oscillating Bodies

The oscillating bodies were two cylinders and a sphere. Cylinders were with different radius to height ratios. Particulars of oscillating bodies are given Table 8.13.

8.4.4.4 Measurements of Physical Parameters

Measured dimensions of the oscillating bodies and other parameters are given in Table 8.14.

8.4.4.5 Temperature Maintenance and Measurement

Rough calculations showed that evaporation of water into dry air would cause un-acceptable cooling of the water surface. To avoid cooling of sample water at the surface, the water saturated air is used to purge the whole system prior to viscosity measurement. The decrement and period of the oscillating system in air is measured in the same water saturated air.

The temperature of air around torsion stand and those parts of the pendulum above the brass tube B was kept higher than that of the lower part by about 0.1 °C.

The temperature was measured by a Pt resistance thermometer by measuring it resistance through a resistance bridge. Uncertainty in resistance measurements leads to uncertainty in temperature to 0.3 mK which is much less than the aimed uncertainty. The indirect extent to which other quantities in the uncertainty budget are affected by a change of 3 mK are all below 0.001 %.

8.4.4.6 Uncertainty Budget

In this case basic sources of uncertainty in measurement of viscosity are

1. Dimensions of oscillating body
2. Damping constant and
3. Angular momentum of the oscillating body

Table 8.14 Measured dimensions of the oscillating bodies and other parameters

Parameter	Cylinder 1	Cylinder 2	Sphere
Radius in mm	9.9862	8.5046	11.9057
Height in mm	20.0055	33.9950	–
Rod radius in mm	0.75	0.75	0.75
Rod immersion in mm	22.0	20.0	17.0
Inertia in 10^{-7} k gm^2	58.772	56.168	65.336
Load coefficient kg^{-1}	3×10^{-3}	3×10^{-3}	3×10^{-3}
Period in water ms	1365.821	1335.313	1440.007
Period in water in ms After buoyancy corrections	1365.834	1335.328	1440.022
Period in air ms	1359.883	1329.608	1433.798
Decrement in water	320.353×10^{-4}	317.030×10^{-4}	318.283×10^{-4}
Decrement in air	17.408×10^{-4}	17.860×10^{-4}	16.582×10^{-4}
Decrement in air of oscillating body	1.866×10^{-4}	1.884×10^{-4}	1.868×10^{-4}
Decrement in liquid of rod	0.260×10^{-4}	0.243×10^{-4}	0.204×10^{-4}
Density of water	998.21	998.20	998.20
Temperature °C	19.988	20.033	20.019
Viscosity	1.00102	0.99998	1.00004

4. Temperature variation of water round the oscillating body.

Measurement standard deviations SD, Relative standard deviation of measurement and its contribution towards uncertainty in viscosity measurement are given in Table 8.15.

The mean viscosity of water at 20 °C and atmospheric pressure is **1.006 ± 0.0005 mPa·s.**

It may be noted that all the authors have used type A method of calculating the various uncertainty components. No consideration has been given due to uncertainty arising by employing type B method. This means that basically uncertainty due to non-repeatability in measurements of various parameters only has been considered.

8.5 Consolidation of Various Viscosity Values

ISO [11] declared the value of dynamic viscosity η for water at 20 °C as 1.0016 mPa·s, and the kinematic viscosity ν of water at 20 °C is taken as 1.0034 $mm^2 s^{-1}$. The estimated uncertainty is 0.17 %. The result is largely based upon the work of Swindells et al. [1], published in 1952,

Bauer et al. of PTB Germany [12] reported the following reasons to reconsider the above statement.

Table 8.15 Uncertainty budget

Parameter	Standard deviation SD	Relative SD %	Contributions to viscosity		
			Cylinder 2	Cylinder 1	Sphere
Radius	0.6 μm	0.006	−0.033	−0.038	−0.035
Height	4 μm	0.020	−0.028	−0.019	–
Moment of Inertia	5×10^{-10} k gm^2	0.009	0.018	0.016	0.014
Net decrement	3×10^{-6}	0.01	0.018	0.018	0.018
Rod immersion	1 mm	4.6	−0.007	−0.007	−0.007
Rod radius	10 μm	1.3	−0.005	−0.005	−0.005
Density of water	0.01 k gm^{-3}	0.001	−0.001	−0.001	0.001
Period in air	2 μs	0.0001	0	0	0
Period in water	25 μs	0.0018	0.002	−0.002	0.003
Temperature	3 mK		0	0	0
Temperature	3 mK		0.007	0.007	0.007
Relative SD of viscosity measurement	–	–	0.051	0.050	0.043

1. According to the ITS 90, 20 °C measured on ITS48 scale is really 19.988 °C, the relative change in viscosity due to 12 mK difference is 0.03 %. This value is significant when repeatability in relative viscosity measurement is 0.01 %.
2. The measurements carried out by Roscoe and Bainbridge [3], Malyarvov [13] and more recent work of Torklep and Oye [3, 14], Kestin and Shankland [4] and Berstad et al. [5] have not been considered.
3. The estimated uncertainty of 0.25 % is unrealistically too high.
4. It is inconsistent to state four valid digits for the dynamic viscosity and five for the kinematic viscosity.
5. Besides 20 °C, reference values of viscosity of high precision are required in a wider range especially at 23 and 25 °C which are frequently used as reference temperatures.

8.5.1 Temperature Dependence

In order to correct viscosity for change in the temperature scale and to determine precise reference values at different temperatures, dependence of viscosity with temperature was evaluated. Relative measurements are much easier and smaller uncertainty is achievable since most sources of systematic deviations are eliminated. Bauer et al. of PTB [16] used results for the viscosity ratios V_r from ten investigations [4, 5, 15–22], which were considered of the highest precision for regression analysis. The temperature range was from 10 to 50 °C. The basic viscometer used in these experiments were capillary in [15–21], oscillating body in [4, 5] and vibrating wire in [22]. It may be noted that V_r is the ratio of the viscosity of water at t °C to that at 20 °C.

Only original data was used in the regression analysis. The viscosity values given in [23] were not independent and were, therefore, not considered. A re-evaluation of data from [19–23] and [24] as published by Kestin et al. [24] using different theoretical working equation was also not considered. The viscosity of steam and water at moderate pressures and temperature was also measured by Moszynski [25]. Of course all data was reduced to IPTS 90 before using for regression purpose.

The temperature dependence is calculated by the least squares fit to the correlation [24] and is given as

$$\log V_r = \frac{(20-t)}{(t+e)} \cdot \left[a + b(20-t) + c(20-t)^2 + d(20-t)^3\right] \tag{8.43}$$

where t is the temperature in °C and five adjustable values are:

$$\begin{aligned} a &= -5.26862 \times 10^{-1} \\ b &= -1.46611 \times 10^{-2} \\ c &= -1.18342 \times 10^{-4} \\ d &= -8.60536 \times 10^{-7} \\ e &= -6.94500 \times 10^{1} \end{aligned} \tag{8.44}$$

The relative standard deviation of (8.44) is 4.60×10^{-4} at 95 % confidence level (coverage factor k = 2).

8.5.2 Pressure Dependence

If water is used as viscosity standard, the pressure dependence of the viscosity value is of minor importance. The influence of pressure on the viscosity is lower by about 3 orders of magnitude than the temperature. Absolute viscosity measurements of the highest precision as well as calibration of viscometers using water are usually performed close to atmospheric pressure. Therefore only viscosity pressure coefficient γ is of interest. Here

$$\gamma = \frac{1}{\eta} \cdot \frac{d\eta}{dp} \tag{8.45}$$

Values of γ for water varies from -6×10^{-10} to 0 per pascal in the temperature range of 10–40 °C.

Table 8.16 Dynamic viscosity values of water

Reference	Year	Measurement temperature °C	Viscosity at 20 °C ITS-90 mPa·s	Reported relative uncertainty
[1] Swindells et al.	1952	20	1.00162	3×10^{-4}
[2] Roscoe et al.	1958	19.75	1.00221	5×10^{-4}
[17] Malyarov	1959	20	1.00334	3×10^{-4}
[4] Kestin et al.	1981	25.6	1.00249	2.6×10^{-3}
[7] Berstad et al.	1988	19.5–25.5	1.00035	5×10^{-4}

Table 8.17 Dynamic and kinematic viscosities of water

S No	t °C	η mPa·s	$\rho \pm 0.005$ k gm^{-3}	ν mm^2s^{-1}	$d\eta/dT$ K^{-1}	$d\nu/dT$ K^{-1}	γ 10^{-4}/mPa
1	15	1.1383 ± 0.0022	999.099	1.1393 ± 0.0022	0.0265	0.0264	−6.14
2	20	1.0020 ± 0.0014	998.203	1.0038 ± 0.0014	0.0245	0.0243	−4.28
3	23	0.9324 ± 0.0017	997.537	0.9347 ± 0.0017	0.0235	0.0232	−3.28
4	25	0.8903 ± 0.0015	997.044	0.8929 ± 0.0016	0.0228	0.0225	−2.65
5	30	0.794 ± 0.0014	995.645	0.8009 ± 0.0014	0.0213	0.0210	−1.22
6	40	0.6527 ± 0.0012	992.212	0.6578 ± 0.0012	0.0188	0.0185	1.20

8.5.3 Mean Value at 20 °C on ITS 90 Scale

The values of dynamic viscosity of water at 20 °C on ITS-90 along with their relative uncertainties are given in Table 8.16.

The value 2.6×10^{-3} for uncertainty reported by Kestin et al. was the sum of the absolute relative uncertainties of input quantities. However it should have been the square root of the sum of squares of the component uncertainties. The correct value should be 1.3×10^{-3}.

There is no evidence that any one investigation is more reliable than others, so equal weight is given to each value and simple mean of 5 values obtained is 1.0020 mPa·s. Three different methods as described in [26, 27] were employed for the calculation of the empirical standard deviations gave the uncertainty for 95 % confidence level as 0.0014 mPa·s. In calculation, it appears that PTB took 1.3×10^{-3} instead of 2.6×10^{-3}.

The viscosity values published by Torklep and Oye in 1979 [3] were obtained with earlier version of the same apparatus. These data are not independent of the results given by Berstad et al. [5], and therefore not taken into account here.

8.5.4 Dynamic, Kinematic Viscosity of Water at Various Temperatures

The values of dynamic viscosity, density, and kinematic viscosity of water along with their temperature coefficients at different temperature, as recommended by PTB Germany, are given in Table 8.17.

References

1. Swindells JF, Coe JR, Godfrey TB (1952) Absolute Viscosity of water at 20 °C. J Res NBS 48:1–31
2. Roscoe R, Bainbridge W (1958) Viscosity determination by the oscillating vessel method II: the viscosity of water at 20 °C. Proc Phys Soc 72:585–595
3. Torklep K, Oye HA (1979) An absolute oscillating cylinder or cup viscometer for high temperatures. J Phys E: Sci Instrum 12:875–885
4. Kestin J, Shankland JR (1981) The Free disk as an absolute viscometer and the viscosity of water in range of 25–150 °C. J Non-Equilib Thermodyn 6:241–256
5. Berstad DA, Knapstad B, Lamvik M, Skjolsvik PA, Torklep K, Oye HA, (1988) Accurate determination of the viscosity of in temperature range 19.5 to 25.5 °C. Physica A, 151:246–280
6. White Hobart S, Kearsley Elliot A (1971) An absolute determination of viscosity using a torsional pendulum. J Res NBS 75A:541–551
7. Kearsley EA (1959) An analysis of an absolute torsional pendulum viscometer. Trans Soc Rheol 111:69–80
8. Barr G (1931) "A monograph of viscometry", Chapter II. Humphrey Milford, New York
9. Roscoe R (1958) Viscosity determination by the oscillating vessel method: theoretical considerations. Proc Phys Soc 72:576–584
10. Kestin J, Paul R, Shankland IR, Khalifa HE (1980) A high temperature, high pressure and oscillating disk viscometer for concentrated ionic solutions. Ber Bunsenges Phys Chem 84:1255
11. ISO Technical Report 3666 (1998) Viscosity of water, 12, Gen Technical Committee ISO/TC 66
12. Bauer H, Binas E, Broeke H, Vokel L (1995) New recommended viscosity values for water as the metrological basis of viscosity. PTB-Mitteilungen 105:99–105
13. Malyarov GA (1959) Determination of viscosity of water at 20 °C. Inst Metrologii 37:125–140
14. Torklep K, Oye HA (1993) Viscosity of water at 20 °C. Int J Thermophys 14:989–991
15. Coe JR, Godfrey TB (1944) Viscosity of water. J Appl Phys 15:625
16. Hardy RC, Cottingen RL (1949) Viscosity of Deuterium oxide and water in the range of 5 to 125. J Res NBS Res 42:573–578 Paper RP 1944
17. Weber W (1955) Viscosity of water from 0 to 40. Z Angew Phys 7:96–98
18. Korosi A, Fabuss BM (1968) Viscosity of liquid water from 25 to 150°. Anal Chem 40:157–162
19. Korson L, Drosi-Hansen W, Millero FJ (1969) J Phys Chem 73:34–39
20. Eicher LD, Zwolinski BJ (1971) Viscosity of super-cooled water and analysis of the extended range temperature coefficient. J Phys Chem 75:2016–2024
21. James CJ, Mulcahy DE, Steel BJ (1984) Viscometer calibration standard_Viscosities of water from 0 to 60 °C and of selected aqueous solutions at 25 °C from measurement with flared capillary viscometer. J Phys D Appl Phys 17:225–230

22. Assael MJ, Polimatidou S, Wakeham WA (1993) The viscosity of liquid water at pressure of 32 Pa. Int J Thermophys 14:225–230
23. Collings AF, Bajenov V (1983) A high precision capillary viscometer and further relative results for the viscosity of water. Metrologia 19:61–66
24. Kestin J, Sokolov M, Wakecham WA (1978) Uncertainty in viscosity of water. J Phys Chem Ref Data 7:941–948
25. Moszynski JR (1961) The viscosity of steam and water at moderate pressures and temperature. J Heat Transf Trans ASME 83:111–124
26. WECC-Doc-19-1990: Western European Calibration Co-operation—"Guidelines for the expression for the uncertainty of measurement in calibration"
27. BS 2846 (1975) Statistical interpretation of data Part 1: Routine analysis of quantitative data. British Standard Institution, London

Appendix A
Standards Pertaining to Viscosity

A.1 ISO Standards for Viscosity Measurement

ISO 2884	Paints and varnishes—Determination of viscosity using rotary viscometers Part 1 Cone and plate viscometers operated at high rate of shear Part 2 Disc or ball viscometers
ISO 3104	Petroleum products—Transparent or opaque liquid—Determination of Kinematic viscosity and calculation of dynamic viscosity
ISO 3105	Glass capillary kinematic viscometers—Specifications and operating instructions
ISO 3219	Plastics—Polymers resins in the liquid state or in emulsions or dispersions—Determination of viscosity using rotational viscometers with defining shear rates
ISO 12058	Plastics—Determination of viscosity using a falling ball viscometer
ISO 1652	Rubber latex—Determination of apparent viscosity by Brookfield test method
ISO 2431	Paints and varnishes—Determination of flow time by use of flow cups
ISO 2555	Plastics—Resins in the liquid state or in emulsions or dispersions—Determination of apparent viscosity by Brookfield method

A.2 ASTM Standards for Viscosity Measurement

ASTM D1200-10	Standard Test Method for Viscosity by Ford Viscosity Cup
ASTM ADJD2161	Viscosity Extrapolation Tables to Zero Degrees Fahrenheit (SSU)
ASTM D2161-10	Standard Practice for Conversion of Kinematic Viscosity to Saybolt Universal Viscosity or to Saybolt Furol Viscosity
ASTM D2162-06	Standard practice for basic calibration of Master viscometers and Viscosity oils
ASTM D2270	Practice for Calculating Viscosity Index from Kinematic Viscosity at 40 and 100 °C
ASTM D2983-09	Standard Test Method for Low-Temperature **Viscosity** of Lubricants Measured by Brookfield Viscometer1, 2
ASTM D3176	
ASTM D341-09	Standard Practice for Viscosity-Temperature Charts for Liquid Petroleum Products

(continued)

S. V. Gupta, *Viscometry for Liquids*, Springer Series in Materials Science 194,
DOI: 10.1007/978-3-319-04858-1,

(continued)

ASTM D3716	Standard test methods for use of emulsion Polymers in Floor Polishers 09-10-1999 (R 2006)
ASTM D4016	Standard test methods for viscosity of chemical grouts by Brookfield Viscometer (Laboratory method
ASTM D4212	Standard test methods for viscosity by dip-type viscosity cups
ASTM D4287-00(2010)	Standard Test Method for High-Shear Viscosity Using a Cone/Plate Viscometer
	1.1 This test method covers the determination of the viscosity of paints, varnishes, and related products at a rate of shear of 12,000 s^{-1}
	1.2 Paints and varnishes that dry very rapidly may not give reproducible results with this test method. Measurements made at elevated temperatures may also give poor precision due to loss of volatiles and to drying.
	1.3 The values stated in SI units
ASTM D445-12	Standard Test Method for Kinematic **Viscosity** of Transparent and Opaque Liquids (and Calculation of Dynamic **Viscosity**)
ASTM D446-07	Standard Specifications and operating instructions for glass Capillary Kinematic Viscometers
ASTM D4741-12	Standard Test Method for Measuring Viscosity at High Temperature and High Shear Rate by Tapered-Plug Viscometer
ASTM D5152	Standard test methods for viscosity of paints and related materials by ISO Flow cups
ASTM D5293	Standard test methods for apparent viscosity of engine oils between −5 and −35 °C
ASTM D5478	Standard test methods for viscosity of materials by falling needle method
ASTM D5481	Standard test methods for measuring apparent viscosity of at high temperature
ASTM D6606-00(2010)	Standard Test Method for Viscosity and Yield of Vehicles and Varnishes by the Duke Viscometer
	This test method covers the procedure for determining the viscosity of varnishes, ink vehicles, and similar liquids that are essentially non volatile and un-reactive under ordinary room conditions using the Duke Automated high-shear rod and collar viscometer
	2.1.2 The instrument in this test method is similar in principle to the falling-rod viscometer described in Test Method D4040
ASTM D7042-12a	Standard Test Method for Dynamic Viscosity and Density of Liquids by Stabinger Viscometer (and the Calculation of Kinematic **Viscosity**)
ASTM D88-07	Standard Test Method for Saybolt Viscosity

A.3 British Standards for Viscosity Measurement

BS 3900-A7.1:2000	Section A7-1: Cone and plate viscometer Methods of test for paints, Tests on liquid paints (excluding chemical tests) Determination of the viscosity of paint at a high rate of shear
BS 3900-A7-2:2003	Section A7-2: Disc or ball viscometer Methods of test for paints, Tests on liquid paints (excluding chemical tests), Determination of the viscosity of paint at a high rate of shear, Disc or ball viscometer operated at a specified speed
BS 188:1977	Methods for determination of the viscosity of liquids
BS EN ISO 1599:1999	Plastics, Cellulose acetate, Determination of viscosity loss on moulding
BS DD CEN/TS 15324:2008	Bitumen and bituminous binders. Determination of equi-viscous temperature based on low shear viscosity using a dynamic shear rheometer in low frequency oscillation mode
BS EN ISO 1599:1999	Plastics. Cellulose acetate. Determination of viscosity loss on moulding
BS ISO 10294-3:1999	Fire-resistance tests. Fire dampers for air distribution systems. Guidance on the test method
BS EN ISO 8619:2004	Plastics, Phenolic resin powder. Determination of flow distance on a heated glass plate

Notes
(1) EN after BS stands for European standard
(2) ISO after BS stands for International Organisation for Standardisation

A.4 Bureau of Indian Standards (BIS) for Viscosity Measurement

IS 1448: Part 56	Methods of Test for Petroleum and its Products—Part 56: Viscosity Index by Calculation	2004
IS 13863	Cereals and milled cereal products—Determination of viscosity of flour—Method using an amylograph	1993
IS 1448: part 146	Methods of test for petroleum and its products—Part 146: Determination of yield stress and viscosity of engine oils at low temperature	1998
IS 1448: Part 25	Methods of Test for Petroleum and its Products—Part 25: Determination of Kinematic and Dynamic Viscosity	1976
IS 244	Method for determination of viscosity (or fluidity) of solutions of cotton and regenerated cellulosic man-made fibres in cuprammonium hydroxide	1984

(continued)

(continued)

IS 11720: Part 3	Methods of test for synthetic rubber—Part 3: Determination of Mooney Viscosity	1993
IS 12869: Part 1	Methods for Determination of Viscosity and Viscometric Fixed Points of Glass—Part 1: Determination of Viscosity by Fibre Elongation Method	1989
IS 12869: Part 2	Methods for determination of viscosity and viscometric fixed points of glass: Determination of softening point (IS 12869: Part 2)	1989
IS 12869	Methods for determination of viscosity and viscometric fixed points of glass—Part 3: Determination of annealing and strain point by beam bending method	1998
IS 13360: Part 11: Sec 10	Plastics—Methods of Testing—Part 11: Special Properties—Section 10: Resins in the Liquid State or as Emulsions or Dispersions—Determination of Apparent Viscosity by the Brookfield Test Method	1999
IS 13360: Part 11: Sec 15	Plastics—Methods of testing—Part 11: Special properties, Section 15 Determination of viscosity using a falling-ball viscometer—inclined-tube method	2006
IS 13360: Part 11: Sec 9	Plastics—Methods of testing—Part 11: Special propertiex, Section 9 Determination of the viscosity of polymers in dilute solution using capillary viscometers—General Principles	2004
IS 8543 : Part 1: Sec 1	Methods of testing plastics—Part 1: Characterization of polymer structure and size, Section 1: Determination of molecular mass from viscosity	1978
IS 4518: Part 1	Methods of test for styrene–butadiene rubbers (SBR)—Part I: Determination of volatile matter, total ash, organic acid, soap, antioxidants, bound styrene and mooney viscosity	1967
IS 6213 : Part 4	Methods of Test for Pulp—Part IV: Determination of Viscosity of Pulp	1971
IS 9466	Viscosity classification of industrial liquid lubricants	1980
IS 9316: Part 2	Methods of test for rubber latex—Part 2: Determination of viscosity (RL:2)	1987
IS 3660: Part 7	Methods of Test for Natural Rubber—Part 7: Determination of Mooney Viscosity 7)	1988

Appendix B
Standard Oils (An Example)

PSL stands for Poulten Selfe and Lee Ltd (PSL)
Russell House, Burnham Business Park
Burnham on Crouch, Essex
CM0 8TE
info@rheotek.com

PSL kinematic viscosity oil standards

PSL Ref	Marked	20 °C	25 °C	40 °C	50 °C	80 °C	100 °C
2700-V01*	N.4	0.47	0.45	0.40	–	–	–
2700-V02*	N.8	0.95	0.89	0.75	–	–	–
2700-V03*	N1.0	1.3	1.2	0.97	0.91	–	–
2700-V04	S3 #	4.6	4.0	2.9	2.4	1.6	1.2
2700-V05	S6	11	8.9	5.7	4.4	2.5	1.8
2700-V06	N10	21	17	10	7.3	4	2.7
2700-V07	S20	44	34	18	13	6	3.8
2700-V08	N35	87	66	32	21	9	5.3
2700-V09	S60	170	120	54	35	12	7.3
2700-V10	N100	330	230	97	60	20	11
2700-V11	S200	660	460	180	105	30	16
2700-V12	N350	1,300	850	310	170	45	23
2700-V13	S600	2,400	1,600	520	280	65	33
2700-V14	N1000	4,690	3,020	940	480	100	48
2700-V15	S2000	9,600	5,900	1,700	890	160	69
2700-V16	N4000	20,000	12,000	3,400	1,600	280	120
2700-V17	S8000	41,000	25,000	6,700	3,200	520	240
2700-V18	N15000	77,000	47,000	13,000	6,100	900	360
2700-V19	S30000	–	89,000	23,000	11,000	1,600	630

(Nominal kinematic viscosity values, stated in mm^2/s, cSt)

S. V. Gupta, *Viscometry for Liquids*, Springer Series in Materials Science 194,
DOI: 10.1007/978-3-319-04858-1, © Springer International Publishing Switzerland 2014

PSL dynamic viscosity oil standards

PSL Ref	Marked	20 °C	25 °C	40 °C	50 °C	80 °C	100 °C
2700-V01*	N.4		0.29	0.26	–	–	–
2700-V02*	N.8	0.73	0.68	0.56	–	–	–
2700-V03*	N1.0	1.0	0.93	0.76	0.71	–	–
2700-V04	S3	3.9	3.3	2.4	1.9	1.2	0.9
2700-V05	S6	9.1	7.6	4.8	3.7	2	1.4
2700-V06	N10	18	15	8.8	6.2	3	2.2
2700-V07	S20	39	31	16	11	5	3.2
2700-V08	N35	76	57	27	18	7	4.3
2700-V09	S60	140	100	46	30	10	6.0
2700-V10	N100	290	210	84	50	17	9.0
2700-V11	S200	580	400	150	90	25	13
2700-V12	N350	1,100	750	270	150	37	19
2700-V13	S600	2,100	1,400	450	240	54	28
2700-V14	N1000	4,080	2,630	810	415	85	39
2700-V15	S2000	7,600	4,900	1,500	770	135	62
2700-V16	N4000	17,000	11,000	2,900	1,400	238	100
2700-V17	S8000	33,000	20,000	5,900	2,800	490	200
2700-V18	N15000	68,000	41,000	11,000	5,400	770	300
2700-V19	S30000	–	80,000	21,000	9,500	1,387	540

(Nominal dynamic viscosity values, stated in mPa·s, cP)

PSL measurement capability

Range	Relative uncertainty Kinematic (mm^2/s)	Relative uncertainty Dynamic (mPa·s)
Up to 6	0.18 %	0.18 %
6–19	0.18 %	0.18 %
19–100	0.22 %	0.22 %
19–100	0.22 %	0.22 %
100–450	0.27 %	0.27 %
450–1,750	0.40 %	0.40 %
1,750 and above	0.66 %	0.66 %

Expanded uncertainty at $k = 2$

Appendix C
Viscosity and Density of Standard Oils

S. V. Gupta, *Viscometry for Liquids*, Springer Series in Materials Science 194,
DOI: 10.1007/978-3-319-04858-1,

Viscosity of some standard oils at different temperature, mm^2/s

Designation	20 °C 68 °F	25 °C 77 °F	37.78 °C 100 °F	40 °C 104 °F	50 °C 122 °F	60 °C 140 °F	80 °C 176 °F	98.89 °C 210 °F	100 °C 212 °F	SUS 100 °F	SUS 210 °F
N.4(b)	0.47	0.45	0.41	0.40	–	–	–	–	–	–	–
N.8(b)	0.74	0.70	0.61	0.60	–	–	–	–	–	–	–
N1.0(b)	1.3	1.2	1.0	0.97	0.91	–	–	–	–	–	–
N2	2.9	2.6	2.1	2.0	1.7	–	1.2	0.95	0.94	–	–
S3(a)	4.6	4.0	3.0	2.9	2.4	–	1.5	1.2	1.2	–	–
N4	6.7	5.8	4.2	4.0	3.2	–	1.9	1.5	1.5	–	–
S6	11	8.9	6.0	5.7	4.4	–	2.4	1.8	1.8	–	–
N7.5	14	12	8.0	7.5	5.8	–	3.1	2.3	2.3	–	–
N10	21	17	11	10	7.3	–	3.5	2.7	2.7	–	–
N14	30	25	15	14	10	–	5.0	3.5	3.4	–	–
S20	44	34	20	18	13	–	5.6	3.9	3.8	100	–
N26	57	46	27	25	18	–	7.9	5.3	5.2	130	–
N35	87	66	35	32	21	–	8.5	5.4	5.3	170	–
N44	110	86	48	44	30	–	12	7.7	7.5	220	–
S60	160	120	60	54	35	–	12	7.7	7.5	280	–
N75	200	150	82	75	50	–	19	12	12	380	–
N100	330	230	110	97	60	–	19	11	11	500	–
N140	400	300	160	140	90	–	31	19	18	720	–
S200	550	400	200	180	110	–	35	22	20	925	105
N250	770	570	280	250	160	–	51	30	29	1,300	140
N350	1,000	720	340	310	190	–	55	34	32	1,570	160
N415	1,400	990	470	415	250	–	77	43	41	2,180	200
S600(c)	1,700	1,300	590	520	310	–	93	51	50	–	240
N750	2,600	1,900	850	750	440	–	130	68	66	–	–
N1000	3,400	2,400	–	940	550	350	150	–	80	–	–

(continued)

(continued)

Designation	20 °C 68 °F	25 °C 77 °F	37.78 °C 100 °F	40 °C 104 °F	50 °C 122 °F	60 °C 140 °F	80 °C 176 °F	98.89 °C 210 °F	100 °C 212 °F	SUS 100 °F	SUS 210 °F
N1400	5,100	3,600	–	1,400	820	510	220	–	120	–	–
S2000	8,300	5,300	1,900	1,600	800	–	160	75	72	–	360
N2500	8,400	6,000	–	2,500	1,500	950	430	–	230	–	–
N4000	20,000	12,000	–	3,400	1,600	850	290	–	120	–	–
N5100	28,000	18,000	–	5,100	2,500	1,300	420	–	170	–	–
S8000	41,000	25,000	8,000	6,700	3,200	–	530	–	240	–	–
N10200	58,000	36,000	–	10,200	4,900	2,500	775	–	300	–	–
N15000	77,000	47,000	–	13,000	6,100	3,000	980	–	360	–	–
N18000	103,000	64,000	–	18,000	8,500	4,300	1,320	–	500	–	–

Density of some standard oils at different temperature, g/cm^3

Designation	20 °C 68 °F	25 °C 77 °F	37.78 °C 100 °F	40 °C 104 °F	50 °C 122 °F	60 °C 140 °F	80 °C 176 °F	98.89 °C 210 °F	100 °C 212 °F
N.4(b)	0.669	0.664	0.652	0.650	–	–	–	–	–
N.8(b)	0.867	0.862	0.851	0.849	–	–	–	–	–
N1.0(b)	0.928	0.861	0.720	0.699	0.616	–	–	–	–
N2	0.762	0.758	0.749	0.748	0.741	–	0.719	0.706	0.705
S3(a)	0.868	0.864	0.855	0.854	0.847	–	0.826	0.812	0.811
N4	0.787	0.783	0.775	0.773	0.767	–	0.746	0.734	0.733
S6	0.878	0.874	0.866	0.864	0.858	–	0.837	0.824	0.823
N7.5	0.801	0.798	0.789	0.788	0.781	–	0.761	0.749	0.748
N10	0.884	0.881	0.872	0.871	0.864	–	0.844	0.832	0.831
N14	0.812	0.809	0.801	0.800	0.793	–	0.774	0.762	0.761
S20	0.863	0.860	0.852	0.850	0.844	–	0.824	0.812	0.811
N26	0.820	0.817	0.809	0.808	0.801	–	0.783	0.771	0.770
N35	0.868	0.864	0.856	0.854	0.849	–	0.830	0.818	0.817
N44	0.828	0.825	0.817	0.816	0.809	–	0.791	0.779	0.778
S60	0.876	0.873	0.865	0.863	0.857	–	0.838	0.826	0.825
N75	0.833	0.830	0.822	0.820	0.814	–	0.796	0.785	0.784
N100	0.882	0.879	0.871	0.870	0.863	–	0.845	0.833	0.832
N140	0.838	0.835	0.827	0.826	0.819	–	0.801	0.790	0.789
S200	0.840	0.837	0.829	0.827	0.821	–	0.803	0.792	0.791
N250	0.842	0.839	0.831	0.830	0.824	–	0.806	0.795	0.794
N350	0.843	0.840	0.833	0.831	0.825	–	0.808	0.797	0.796
N415	0.846	0.843	0.835	0.834	0.828	–	0.810	0.799	0.798
S600(c)	0.847	0.844	0.836	0.835	0.829	–	0.811	0.800	0.799
N750	0.849	0.846	0.838	0.837	0.831	–	0.813	0.802	0.801
N1000	0.850	0.847	–	0.838	0.832	0.826	0.814	–	0.802
N1400	0.851	0.848	–	0.839	0.833	0.827	0.815	–	0.804
S2000	0.877	0.874	0.867	0.865	0.860	–	0.842	0.832	0.831
N2500	0.849	0.846	–	0.837	0.832	0.826	0.814	–	0.802
N4000	0.883	0.880	–	0.872	0.866	0.860	0.850	–	0.838
N5100	0.886	0.883	–	0.875	0.870	0.864	0.853	–	0.842
S8000	0.890	0.888	0.881	0.879	0.874	–	0.858	–	0.847
N10200	0.891	0.888	–	0.880	0.874	0.869	0.858	–	0.847
N15000	0.894	0.891	–	0.823	0.877	0.872	0.861	–	0.850
N18000	0.895	0.892	–	0.884	0.879	0.873	0.863	–	0.852
S30000	–	0.894	0.887	0.886	0.881	–	0.865	–	0.854

Appendix D
Buoyancy Correction

Data for buoyancy and temperature corrections

Viscosity standards	Kinematic viscosity mm^2/s		Density g/ml		$\rho_{n-1} - \rho_n$ at 40 °C
	37.78 °C	40 °C	20 °C	40 °C	
S3(a)	3.0	2.9	0.868	0.748	–
N10	11	10	0.884	0.788	0.040
N35	35	32	0.868	0.808	0.020
N100	110	97	0.882	0.820	0.012
N350	340	310	0.843	0.830	0.008
N1000	550**	940	0.850	0.837	0.007
N2500	1,500**	2,500	0.849	0.865	0.028
S8000	8,000	6,700	0.890	0.875	0.010
N10200	4,900**	10,200	0.891	0.879	0.004
S30000	28,000	23,000	–	0.884	0.005

D.1 Buoyancy Correction Water at 20 °C to Oil at 40 °C

Buoyancy correction to be applied to the viscometer constant of the first viscometer which was calibrated taking water as primary standard at 20 °C but is used to determine the viscosity of the oil at 40 °C is calculated as follows:

Density of water at 20 °C = 998.4081 at 40 °C = 992.2166 kg/m^3
Density of air at 20 °C = 1.199217 at 40 °C = 1.111770 kg/m^3

Relative Buoyancy correction c_b to viscometer constant for the viscometer calibrated by water at 20 °C but is used to determine the viscosity of oil at 40 °C

$$c_b = \frac{\sigma_{20}}{\rho_{20}} - \frac{\sigma_{40}}{\rho_{o40}}$$

Here ρ_{o40} is the density of first oil to be used with this viscometer

S. V. Gupta, *Viscometry for Liquids*, Springer Series in Materials Science 194,
DOI: 10.1007/978-3-319-04858-1,

$$\begin{aligned} c_b &= 1.199217/998.4081 - 1.111770/748 \\ &= 0.00120\,113 - 0.00148632 \\ &= -0.000\,285 \\ &\cong -0.03\,\%. \end{aligned}$$

D.1 Maximum Buoyancy Correction Oil to Oil at 40 °C

Relative buoyancy corrections defined applicable to consecutive viscometers

$$c_b = \sigma_{40}(\rho_{n-1} - \rho_n)/\rho_n\rho_{n-1}$$

We see that maximum difference between the densities of any two consequent oils is 0.044 kg/m^3. Hence maximum relative buoyancy correction for a viscometer used at 40 °C with any oil

$c_b = 1.111770(1/748 - 1/788) = 0.0013368 - 0.00126903 = 0.000065$ or 0.0065 % which is negligible in comparison to the target uncertainty of 0.1 %.

Appendix E
Coefficients of Viscosity of Some Standard Oils

Kinematic viscosity of oils is temperature dependent. Coefficient of increase in kinematic viscosity is defined as

$$\alpha_v = (\nu_{40} - \nu_{38.78})/\nu_{40}$$

Viscosity of some oils at 38.78 °C (100 °F) and at 40 °C

Viscosity standard	Kinematic viscosity		Coeff. of viscosity
	38.78 °C	40 °C	
S3(a)	3.0	2.9	0.0155
N10	11	10	0.045
N35	35	32	0.0422
N100	110	97	0.0604
N350	340	310	0.0436
N1000	550**	940	0.0415
N2500	1,500**	2,500	0.0400
S8000	8,000	6,700	0.0873
N10200	4,900**	10,200	0.1082
S30000	28,000	23,000	0.0979

Coefficients of expansion of standard oils are given in column 4 of the table.

S. V. Gupta, *Viscometry for Liquids*, Springer Series in Materials Science 194,
DOI: 10.1007/978-3-319-04858-1, © Springer International Publishing Switzerland 2014

Appendix F
Equivalent Viscosities at 100 °F

F.1 Lower Range

Approximates values of kinematic viscosity on different scales corresponding to the values in mm^2/s, a sub-multiple SI unit of kinematic viscosity

Kinematic viscosity (mm^2/s)	Saybolt universal seconds (SUS)	ASTM viscosity grade +10 % SUS*	Saybolt Furol seconds	Engler degrees	Redwood No. 1 seconds	Redwood admiralty No. 2 seconds
2	32.62	32	–	1.1195	–	–
4	39.14	–	–	1.3075	36.5	–
6	45.56	40	–	1.4805	41.9	–
8	52.09	–	–	1.6535	47.6	–
10	58.91	60	–	1.834	53.5	–
12	66.04	–	–	2.023	59.6	–
14	73.57	75	–	2.222	66.1	–
16	81.30	–	–	2.435	72.7	–
18	89.44	–	–	2.646	79.6	–
20	97.77	–	–	2.876	86.6	–
22	106.4	105	–	3.11	93.9	–
24	115.0	–	–	3.35	101.3	–
26	123.7	–	–	3.59	108.8	–
28	132.7	–	–	3.83	116.4	–
30	141.3	–	–	4.08	124.0	–
32	150.2	150	–	4.33	1319	13.9
34	159.2	–	–	4.58	1.39.9	–
36	168.2	–	–	4.84	147.6	(16.2)
38	177.3	–	–	5.09	153.6	(16.9)
40	186.3	–	–	5.35	163.5	–

Notes (1) () Figures in parentheses means approximate
(2) Tentative nominal viscosity grade at 1,000 °F. Each grade permits a maximum range of ± 10% of the nominal viscosity
(3) Centipoise are obtained by multiplying kinematic viscosity in centistokes by specific gravity
(4) 1 centipoise symbol (cP) = 1 millipascal·second symbol (mPa·s)
(5) 1 centistoke symbol (cSt) = 1 millimetre square per second symbol (mm^2 /s)

S. V. Gupta, *Viscometry for Liquids*, Springer Series in Materials Science 194,
DOI: 10.1007/978-3-319-04858-1,

F.2 Higher Range

Kinematic viscosity (mm^2/s)	Saybolt universal seconds (SUS)	Viscosity grade ±10 % SUS*	Saybolt Furol seconds	Engler degrees	Redwood standard No. 1 seconds	Redwood admiralty No. 2 seconds
42	195.3	–	–	5.61	171.6	–
44	204.4	–	–	5.87	179.7	–
46	213.7	215	–	6.12	187.7	(20.2)
48	222.9	–	–	6,38	195.7	(20.7)
50	232.1	–	26.3	6,64	203.8	(21.4)
55	255.2	–	28.72	7.30	224.1	(23.4)
60	278.3	–	30.9	7.95	244.5	(25.6)
65	301.4	315	32.9	8.61	264.7	(27.8)
70	324.4	–	35.0	9.26	286.1	(30.0)
75	347.6	–	37.5	9.92	306	31.8
80	370.8	–	40.0	10.57	326	33.8
85	393.9	–	42.0	11.23	346	35.8
90	417.1	–	44.3	11.89	367	37.7
95	440.3	–	46.5	12.54	387	39.7
100	463.5	465	48.5	13.20	408	41.7
110	509.8	–	54	14.52	449	45.7
120	556.2	–	58	15.84	490	49.6
130	602.5	–	63	17.16	530	53.7
140	648.9	–	67	18.48	571	57.7
150	695.2	700	72	19.80	612	61.7
160	741.6	–	77	21.12	653	65.7
170	787.9	–	82	22.44	707	69.8
180	834.2	–	86	23.76	734	73.8
190	880.6	–	90	25.08	775	77.9
200	926.9	–	95	26.40	816	81.9
220	1019.6	1,000	104	29.04	898	90.0
240	1112.3	–	114	31.68	979	98.1
260	1205.0	–	123	34.32	1,061	106.2
280	1297.7	–	1333	36.96	1,142	114.4
300	1390.4	–	143	39.60	1,224	122.5
320	1483.1	1,500	153	42.24	1,306	130.7
370	1714.8	–	175	48.84	1,510	–
464	2155.1	2,150	220	61.25	1,893	–
500	2317.4	–	238	66.00	2,040	204.3
580	3150.0	3,150	(325)	89.76	2,774	(280)
(1,000)	4,650	4,650	(465)	(133)	(4,050)	–
1,510	7,000	7,000	(700)	199.3	616	–

Appendix G
Moment of Inertia

Assumptions:

(1) Every solid body is made of materials of uniform density and has a mass denoted by m.
(2) A particle of mass m has no dimensions (length, breadth or height).

G.1 Particle

The moment of inertia I of a particle of mass m and at a distance r from the axis of rotation is given as

$$I = mr^2.$$

G.2 Rod of Length L of Circular Section of Radius r

G2.1 The moment of inertia I about the axis passing through its mid point Fig. G.1 is given as

$$I = \frac{mL^2}{12}$$

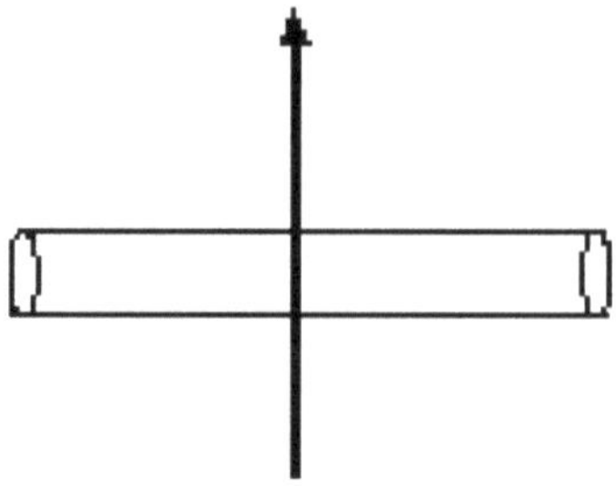

Fig. G.1 *Rod* of *circular* section

DOI: 10.1007/978-3-319-04858-1,

G2.2 The moment of inertia I about the axis passing through its either end Fig. G.2 is given as

$$I = \frac{mL^2}{3}$$

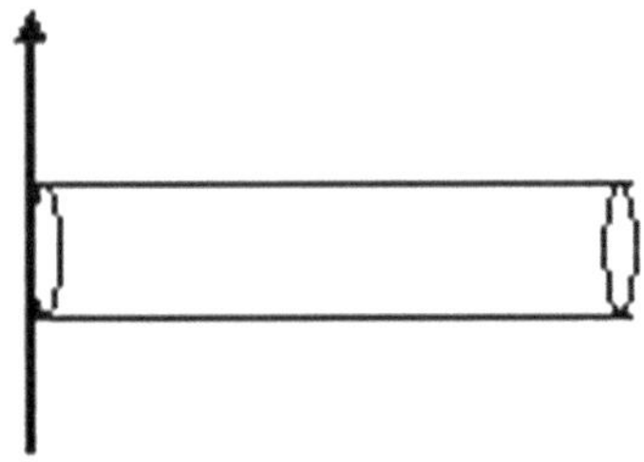

Fig. G.2 *Rod* of *circular* section

G.3 Circular Plate of Radius r

G3.1 The moment of inertia I_z about z axis (axis normal to the plate and passing through its centre Fig. G.3 is given as

$$I_z = \frac{mr^2}{2}$$

G3.2 The moment of inertia I about x or y axis (any diameter of the plate) is given as

$$I = \frac{mr^2}{4}$$

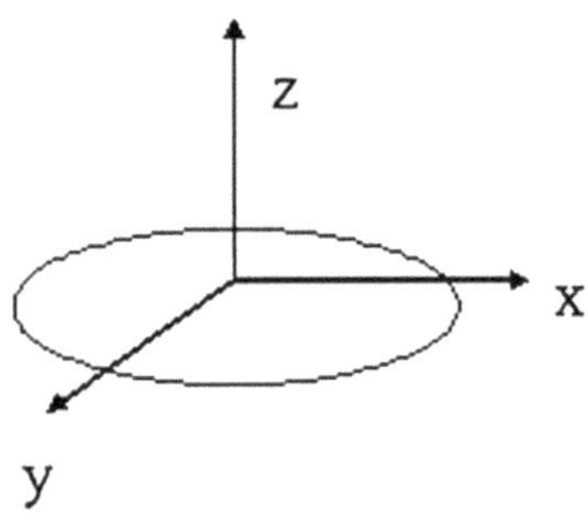

Fig. G.3 *Circular disc*

G.4 A Rectangular Plate of Uniform Thickness, Length a and Breadth b

G4.1 The moment of inertia *I* about the axis passing through its centre Fig. G.4 is given as

$$I = m\frac{a^2 + b^2}{12}$$

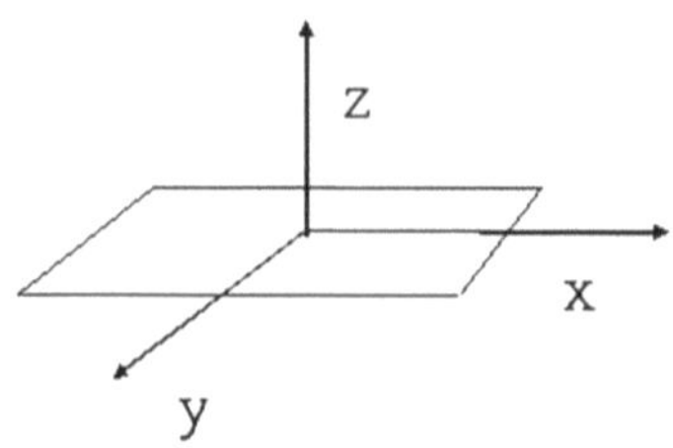

Fig. G.4 *Rectangular* plate

G4.2 The moment of inertia *I* about the axis passing through its corner Fig. G.5 is given as

$$I = m\frac{a^2 + b^2}{3}$$

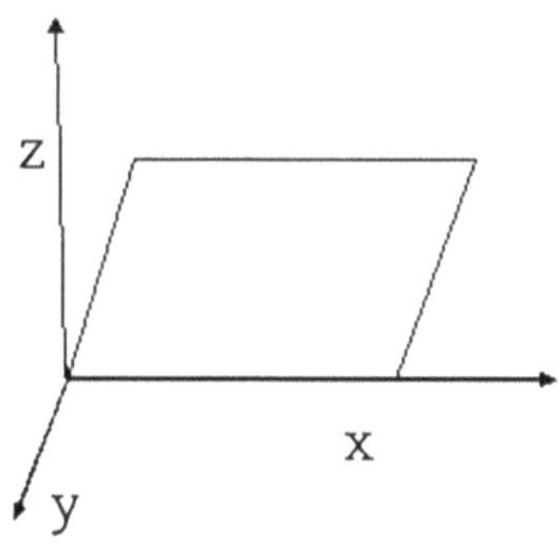

Fig. G.5 *Rectangular* plate

G.5 Thin Cylindrical Shell of Uniform Thickness

The moment of inertia I about the axis of a cylindrical shell Fig. G.6 is given as

$$I = mr^2$$

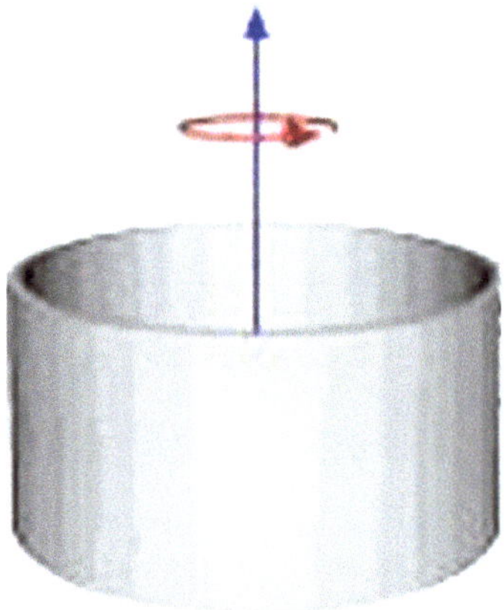

Fig. G.6 *Cylindrical* shell

G.6 Solid Cylinder of Radius r and Height h

G6.1 The moment of inertia I of a solid cylinder about its axis (z axis) Fig. G.7 is given as

$$I = \frac{mr^2}{2}$$

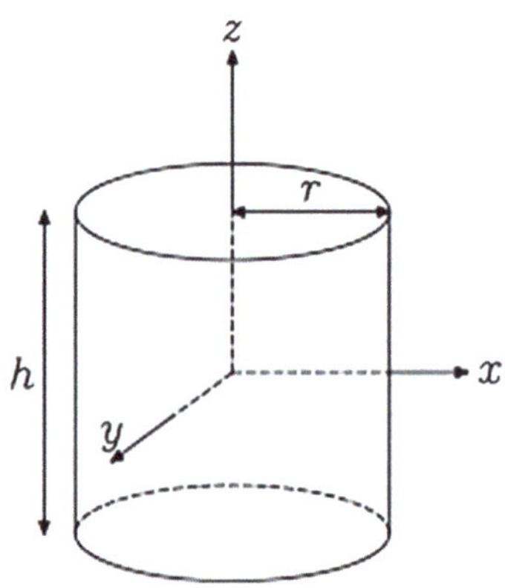

Fig. G.7 *Solid cylinder*

G6.2 The moment of inertia I of the solid cylinder about x or y axis Fig. G.7 is given as

$$I = \frac{1}{12}\left(3r^2 + h^2\right)$$

G.7 Solid Hollow Cylinder of Inner Radius r_1 and Outer Radius r_2

G7.1 The moment of inertia I_z of a hollow cylinder Fig. G.8 about its axis (z axis) is given as

$$I_z = \frac{1}{2}m\left(r_2^2 + r_1^2\right)$$

G7.2 The moment of inertia of hollow cylinder about x or y axis is given as

$$I_x = I_y = \frac{1}{12}\left\{3\left(r_1^2 + r_2^2\right) + h^2\right\}$$

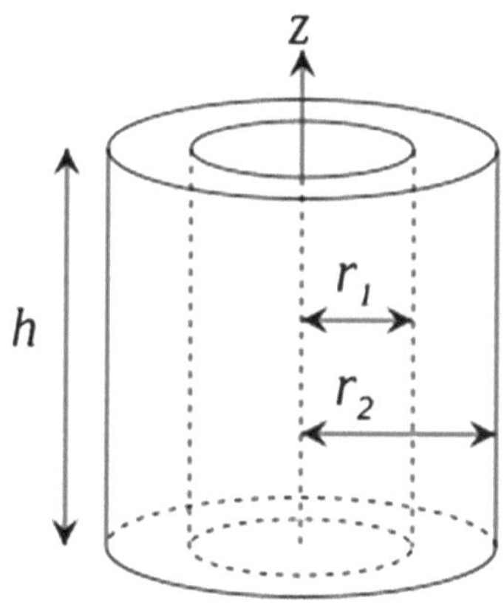

Fig. G.8 *Hollow cylinder*

G.8 Solid Sphere or Radius r

The moment of inertia I of a solid sphere about any of its diameter as axis Fig. G.9 is given as

$$I = \frac{2mr^2}{5}$$

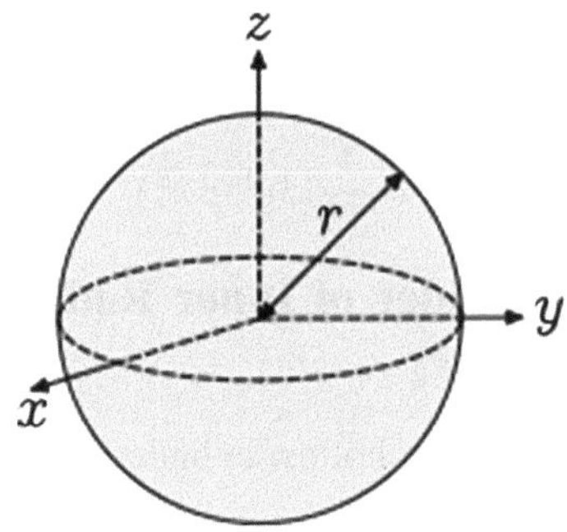

Fig. G.9 *Solid sphere*

G.9 Spherical Shell of Radius r

The moment of inertia I of a spherical shell about any of its diameter Fig. G.10 is given as

$$I = \frac{2mr^2}{3}$$

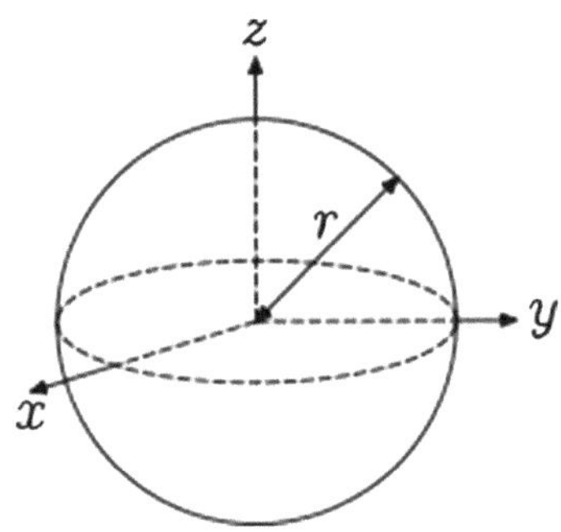

Fig. G.10 *Spherical* shell

G.10 Hollow Sphere Inner Radius r_1 and Outer Radius r_2

The moment of inertia I of a hollow sphere of about its any diameter Fig. G.11 is given as

$$I = \frac{2m\left(r_2^5 - r_1^5\right)}{5\left(r_2^3 - r_1^3\right)}$$

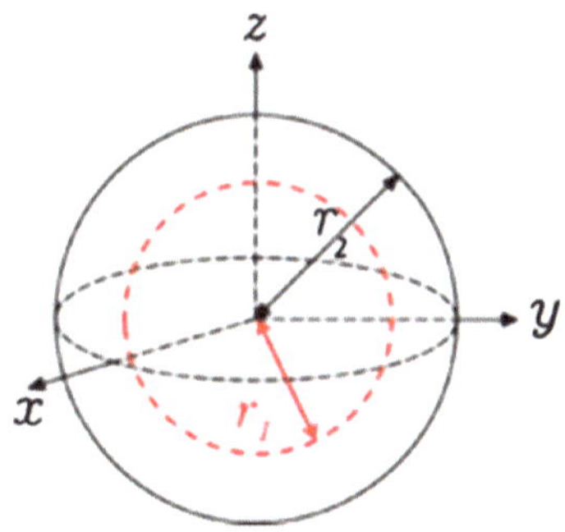

Fig. G.11 *Hollow sphere*

G.11 Solid Vertical Cone of Radius r and Height h

G11.1 The moment of inertia I_z of right circular solid cone Fig. G.12, about z axis is given as

$$I_z = \frac{3}{10} m r^2$$

G11.2 The moment of inertia of right circular solid cone about x or y axis Fig. G.10 is given as

$$I = \frac{3}{4} m \left(\frac{r^2}{2} + h^2 \right)$$

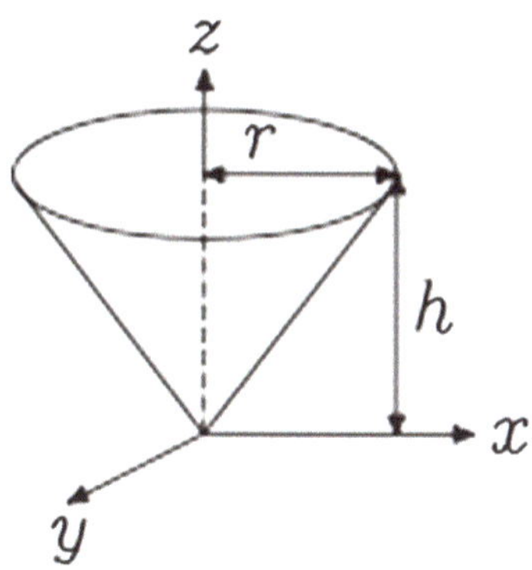

Fig. G.12 *Solid circular cone*

G.12 Rectangular Plate of Breadth b and Length a

The moment inertia of a rectangular plate about axis normal to its plane at the edge of its middle point, Fig. G.13, is given as

$$I = \frac{ma^2}{3} + \frac{mb^2}{12}$$

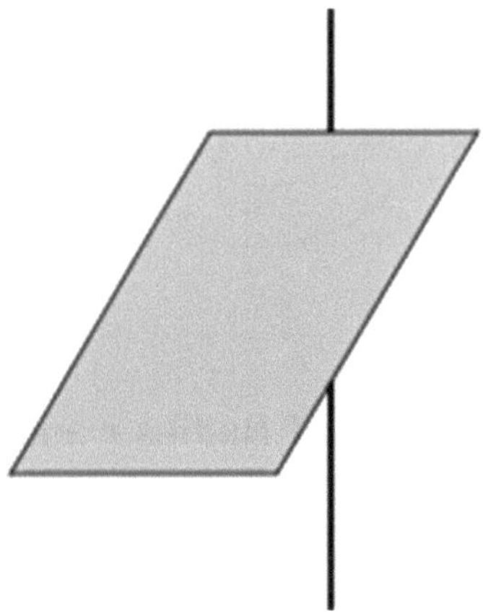

Fig. G.13 *Rectangular* plate

G.13 Rectangular Block of Length a, Breadth b and Height c

G13.1 Moment of inertia I_c of a solid rectangular block about the axis passing through it centre of gravity along its height c, Fig. G.12, is given as

$$I_c = \frac{m(a^2 + b^2)}{12}$$

G13.2 Moment of inertia I_a of a solid rectangular block about the axis passing through it centre of gravity along its length a is

$$I_a = \frac{m(b^2 + c^2)}{12}$$

G13.3 Moment of inertia I_b of a solid rectangular block about the axis passing through it centre of gravity along its breadth b is (Fig. G.14)

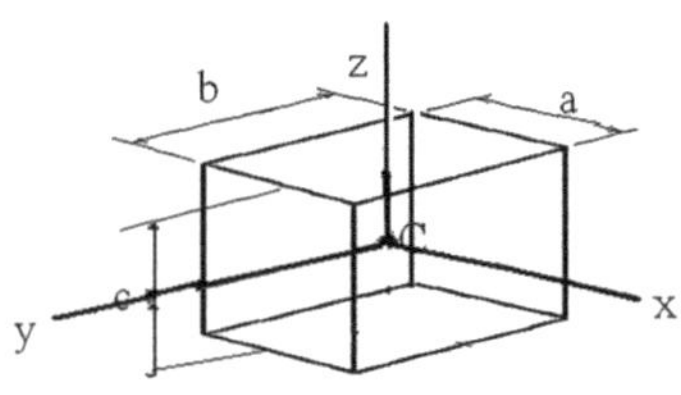

Fig. G.14 *Rectangular* block

$$I_b = \frac{m(c^2 + a^2)}{12}$$

G13.4 Moment of inertia I of a solid rectangular block about its longest diagonal, Fig. G.15, is given as

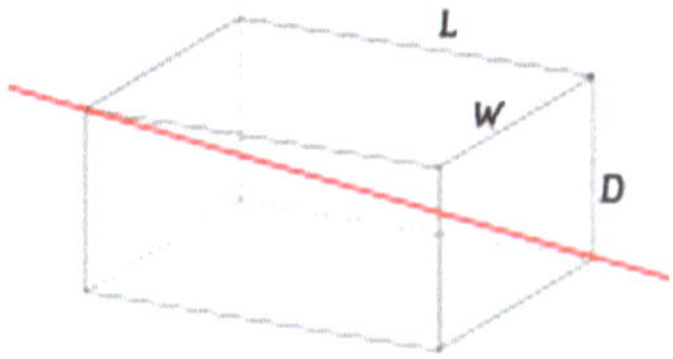

Fig. G.15 *Solid rectangular* block

$$I = \frac{m(a^2b^2 + b^2c^2 + c^2a^2)}{6(a^2 + b^2 + c^2)}$$

References

1. Serway RA (1986) Physics for scientists and engineers, 2nd edn. Saunders College Publishing, p 202, ISBN 0-03-004534-7
2. Classical Mechanics—Moment of inertia of a uniform hollow cylinder. www.LivePhysics.com. Accessed 31 Jan 2008
3. Beer FP, Russell Johnston E Jr (1984) Vector mechanics for engineers, 4th edn. McGraw-Hill, New York, p 911, ISBN 0-07-004389-2
4. Weisstein EW (2010) Moment of inertia—ring. Wolfram Research. Accessed 25 March 2010

Index

S. V. Gupta, *Viscometry for Liquids*, Springer Series in Materials Science 194,
DOI: 10.1007/978-3-319-04858-1, © Springer International Publishing Switzerland 2014

MIX
Papier aus verantwortungsvollen Quellen
Paper from responsible sources
FSC® C105338

If you have any concerns about our products,
you can contact us on
ProductSafety@springernature.com

In case Publisher is established outside the EU,
the EU authorized representative is:
Springer Nature Customer Service Center GmbH
Europaplatz 3, 69115 Heidelberg, Germany

Printed by Libri Plureos GmbH
in Hamburg, Germany